Nuclear Terrorism and National Preparedness

NATO Science for Peace and Security Series

This Series presents the results of scientific meetings supported under the NATO Programme: Science for Peace and Security (SPS).

The NATO SPS Programme supports meetings in the following Key Priority areas: (1) Defence Against Terrorism; (2) Countering other Threats to Security and (3) NATO, Partner and Mediterranean Dialogue Country Priorities. The types of meeting supported are generally "Advanced Study Institutes" and "Advanced Research Workshops". The NATO SPS Series collects together the results of these meetings. The meetings are co-organized by scientists from NATO countries and scientists from NATO's "Partner" or "Mediterranean Dialogue" countries. The observations and recommendations made at the meetings, as well as the contents of the volumes in the Series, reflect those of participants and contributors only; they should not necessarily be regarded as reflecting NATO views or policy.

Advanced Study Institutes (ASI) are high-level tutorial courses to convey the latest developments in a subject to an advanced-level audience

Advanced Research Workshops (ARW) are expert meetings where an intense but informal exchange of views at the frontiers of a subject aims at identifying directions for future action

Following a transformation of the programme in 2006 the Series has been re-named and re-organised. Recent volumes on topics not related to security, which result from meetings supported under the programme earlier, may be found in the NATO Science Series.

The Series is published by IOS Press, Amsterdam, and Springer, Dordrecht, in conjunction with the NATO Emerging Security Challenges Division.

Sub-Series

A.	Chemistry and Biology	Springer
B.	Physics and Biophysics	Springer
C.	Environmental Security	Springer
D.	Information and Communication Security	IOS Press
E.	Human and Societal Dynamics	IOS Press

http://www.nato.int/science
http://www.springer.com
http://www.iospress.nl

Series B: Physics and Biophysics

Nuclear Terrorism and National Preparedness

edited by

Samuel Apikyan
National Center of Security and Protection, USA

and

David Diamond
Brookhaven National Laboratory, USA

Springer

Published in Cooperation with NATO Emerging Security Challenges Division

Proceedings of the NATO Advanced Research Workshop on
Preparedness for Nuclear and Radiological Threats
Los Angeles, USA
18–20 November 2014

Library of Congress Control Number: 2015944111

ISBN 978-94-017-9935-5 (PB)
ISBN 978-94-017-9890-7 (HB)
ISBN 978-94-017-9891-4 (e-book)
DOI 10.1007/978-94-017-9891-4

Published by Springer,
P.O. Box 17, 3300 AA Dordrecht, The Netherlands.

www.springer.com

Printed on acid-free paper

Preface

The world faces no greater or more urgent danger than a terrorist attack with the intent of killing, maiming, and traumatizing a large population. International peace and security is threatened in particular by the proliferation of nuclear materials and technologies that could lead to a nuclear or radiological attack. More nations are trying to acquire nuclear weapons, and black markets trade in nuclear secrets and materials. Terrorists are determined to buy, build, or steal a nuclear weapon or use a radioactive source in a conventional bomb.

Organizations like al Qaeda and the so-called Islamic State have said that obtaining these weapons and perpetrating another "Hiroshima" are their "religious duty." Organizations such as these have the will, the technical know-how, and the financial resources to make these threats a reality.

Our strategy to combat these threats is multilayered, and events in recent years have shown the necessity to continually reevaluate national preparedness programs. Throughout the world there are people working on the key issues related to this subject such as:

- Preventing, avoiding, or stopping threats
- Protecting our citizens and assets against the greatest threats and hazards
- Mitigating the loss of life and property by lessening the impact of future disasters
- Responding quickly to save lives, protect property and the environment, and meet basic human needs in the aftermath of a catastrophic incident
- Recovering through timely restoration and strengthening of infrastructure and the economy, as well as the social fabric of communities affected by a catastrophic incident

The NATO Advanced Research Workshop on "Preparedness for Nuclear and Radiological Threats" was held in Los Angeles, on 18–20 November 2014 with support from the NATO Science for Peace and Security Programme. The purpose of the workshop was to contribute to the critical assessment of existing knowledge on this subject, to identify directions for future research and policies, and to promote close working relationships between scientists, engineers, and policy makers from different countries and with different professional experience. More than 100 representatives of 18 countries participated. The program was built upon the accomplishments of The Hague 2014 Nuclear Security Summit and previous NATO workshops such as "Countering Nuclear/Radiological Terrorism" (2005); "Prevention, Detection and Response to Nuclear and Radiological Threat" (2007); and "Threat Detection, Response and Consequence Management Associated with Nuclear and Radiological Terrorism" (2008).

This book contains approximately half of the papers presented at the workshop. The other half of the papers are found in the book *Nuclear Threats and Security Challenges*. We hope it will be useful not only for the multinational scientific and technical communities engaged in combating nuclear and radiological terrorism but also for decision makers and for those working at governmental and policy levels whose actions affect the directions the science takes and how the technology is incorporated into country-specific national systems for combating nuclear and radiological threats.

Los Angeles — Samuel Apikyan
Upton — David Diamond

Contents

Part II Detection and Protection

Part III Prevention and Mitigation

Part I
Response and Recovery

Chapter 1
Preparing for the Aftermath of a Nuclear Detonation; An Analytic Framework for Disaster Management

Brooke Buddemeier and Nancy Suski

Abstract Preparedness activities for complex, technical hazards require a sound scientific basis integrated into appropriate guidelines and preparedness activities. The Federal Planning Guidance for Response to a Nuclear Detonation provides the strategy for response to an improvised nuclear device detonation and was built on an analytical framework of supporting science developed by national laboratories and other technical organizations. Recent advances in our understanding of the hazards posed by such an event includes detailed fallout predictions from the advanced suite of three-dimensional meteorology and plume/fallout models developed at Lawrence Livermore National Laboratory, including extensive global geographical and real-time meteorological databases to support model calculations.

This is an updated case study of the analytic framework for disaster management being applied to response preparedness. The methodology and results, including visualization aids developed for response organizations, have greatly enhanced the community planning process through first-person points of view and description of the dynamic nature of the event.

1.1 Introduction

The Federal Emergency Management Agency (FEMA) provides technical guidance for regional, state and local responders who have responsibility for developing local Improvised Nuclear Device (IND) response plans. In support of these preparedness activities, Lawrence Livermore National Laboratory (LLNL) provides advanced modeling; technical assessments; briefings; and reports to inform Federal, state, and local response and recovery planning activities. This technical work provides the analytic framework for sound Federal, state, local, and private sector nuclear terrorism response planning.

B. Buddemeier (✉) • N. Suski
Lawrence Livermore National Laboratory, Livermore, CA, USA
e-mail: buddemeier1@llnl.gov

S. Apikyan, D. Diamond (eds.), *Nuclear Terrorism and National Preparedness*, NATO Science for Peace and Security Series B: Physics and Biophysics, DOI 10.1007/978-94-017-9891-4_1

This document provides an overview of the potential impacts of a nuclear detonation, and the planning considerations that could dramatically reduce the number of casualties by helping the public and responders avoid exposure and support rapid response.

Extensive review of nuclear weapon effects studies and discussions with nuclear weapon effects experts from various federal agencies, national laboratories, and technical organizations have identified key issues and bounded many of the unknowns required to support response planning for a low-yield, ground-level nuclear detonation in a modern U.S. city.

The scale and severity of disasters are growing and will likely pose systemic threats that require an engaged and resilient community to address [1]. Disaster-management strategies will be affected by many factors, including global interdependencies, technical innovation and dependency, the evolving terrorist threat, and the changing role of the individual. Meeting these challenges requires much more sophisticated methodologies and innovative approaches to predicting, understanding, and mitigating potential hazards through the development of effective response plans to improve resiliency.

Preparing for a catastrophic event is quantitatively different in terms of mass casualties and fatalities, infrastructure damage, and disruption of life-sustaining services. For complex technical threats involving weapons of mass destruction (WMD), actions taken now to understand and plan for the immediate aftermath of such an event will be critical to saving lives and rebuilding communities. The Disaster Management Analytic Framework (DMAF) [2] enables emergency planners to quantify and visualize the impact of a significant WMD incident. DMAF provides insights that can lead to more effective preparedness for response and recovery operations and can be applied to other catastrophic planning activities.

Scenario #1 of the 15 Department of Homeland Security (DHS) national planning scenarios is a 10 kt improvised nuclear device (IND) detonation. An effective response involves managing a large-scale incident, including mass casualty, mass evacuation, and mass decontamination issues. Preparedness planning activities based on the nuclear terrorism scenario present difficult challenges in time-critical decision making; the need to coordinate large-scale response operations across multiple jurisdictions; and the need to effectively respond with limited infrastructure and resources. A DMAF for scenario #1 was utilized to define key characteristics of the event and aid response planning.

In 2007, the U. S. Congress expressed concern that cities have little guidance to help them prepare their populations for the critical moments shortly after a nuclear terrorism event. They directed the DHS, Office of Health Affairs (OHA) to work with the National Academies Institute of Medicine, the Homeland Security Institute, the national laboratories, and state and local response organizations to address this issue [3]. The OHA initiative is currently managed by FEMA as part of a coordinated federal effort to improve response planning for a nuclear detonation.

1.2 Need for Planning Guidance

Federal protective action guidance for radiation exposure has existed for decades; however, the focus has been concentrated on avoiding relatively low-level exposures to decrease the risk of cancer from an accidental transportation or nuclear power plant release. The 2008 Federal Register Notice published by DHS [4], which clarified how existing protective action guidance can be applied for radiological and nuclear terrorism, did not specifically address guidance for the acute effects of a domestic nuclear explosion. This was a recognized deficiency and the analytical framework approach was used provide supporting science to inform a development effort by the Office of Science and Technology Policy (OSTP) to provide guidance for responding to the aftermath of a nuclear detonation in a modern city [5]. A 2013 update of the protective action guides [6] reaffirmed that "PAGs were inadequate for early response planning needs specific to an IND" and the OSTP guidance is the appropriate document for this information.

The Cold War civil defense program can help with some insights and advice, but many of the paradigms no longer apply. For example, the concept of a fallout shelter worked well with a few minutes warning of incoming missiles, but its applicability is less clear for an attack that occurs without any notice. There also appeared to be a lack of scientific consensus on the appropriate actions to take after a nuclear detonation. For example, the recommendations of DHS's Ready.gov, which are consistent with the recommendations of the National Academy of Sciences [7], were recently criticized by the Federation of American Scientists [8] because of conflicting recommendations with a RAND study [9].

The work presented here demonstrates how the DMAF principle was used to update the Cold War guidance to address the asymmetric threat we now face. Both society and urban environments have changed significantly over the last half-century, and new preparedness guidance is required.

1.3 Methodology

This DMAF methodology provided key planning considerations and response strategies associated with response to a nuclear detonation. These strategies are designed to protect response personnel, provide regional situational assessment, and support public health and safety. A capabilities-based planning approach utilized extensive modeling and technical analysis and discussions with almost 500 emergency responders from across the nation to develop key response planning factors. The DMAF provides a common foundation that facilitates the development of strategic response priorities and enables a more collaborative, transparent, and responsive analysis for planning scenarios. Every community has unique requirements and may reasonably adopt different response strategies based on the same analysis. For example, the importance of early, adequate shelter followed by informed evacuation as a key public protection strategy will be applied differently in a community that lacks an abundance of adequate shelters or effective evacuation routes.

To resolve conflicts in the technical community and create a coordination point for research, DHS formed the IND Modeling and Analysis Coordination Working Group (MACWG). Membership includes national laboratories, technical organizations, and federal agencies. The purpose of the MACWG is to establish scientific consensus (where possible) on nuclear weapons effects; bound uncertainties and identify unknowns; and resolve conflicts with respect to recommended response actions. The MACWG brought together the collective capabilities in nuclear-weapon effects modeling, atmospheric transport and dispersion, radiation health physics, and blast and shelter analyses to create a scientific basis for IND response planning. The MACWG has enabled better coordination within the federal, state, and local community and has provided a venue in which to discuss critical scientific and technical issues that must be resolved to save lives and ensure resilience to disasters.

1.4 Nuclear Detonation Effects

The basic anatomy of a nuclear explosion is well known and documented in literature such as Glasstone's The Effects of Nuclear Weapons [10] and NATO documents [11]. Mitigating the impact of a domestic nuclear explosion requires a basic understanding of key effects. These effects can be broken into two main components: prompt and delayed. As an example, the effects identified below are approximate for a ten kiloton (10 kt) nuclear explosion in a generic large city. This is consistent with the national planning scenario #1 and with early nuclear weapons such as those used on Hiroshima and Nagasaki.

Primary among prompt effects is blast (Fig. 1.1). A 10 kt explosion is equivalent to 5,000 truck bombs like the one used to destroy the Murrah building in the 1995 Oklahoma City bombing [12]. Blast will damage or destroy most buildings within a half-mile of the detonation location, and it is unlikely that the population in this area would survive. From a half-mile to about a mile out, survival will mostly likely depend on the type of structure a person was in when the blast occurred. Even at a mile, the blast wave will have enough energy to overturn some cars and severely damage some light structures.

A mile from the detonation is also the approximate distance at which a person outdoors could receive a significant exposure of initial ionizing radiation. The closer to the detonation point, the higher the exposure. The same is also true for an outdoor individual's exposure to the thermal pulse from the detonation, which may also cause burns to exposed skin out to this range, and possibly further on a day with good visibility. Both of these effects are reduced for people inside buildings or in the shadow of buildings in the urban area.

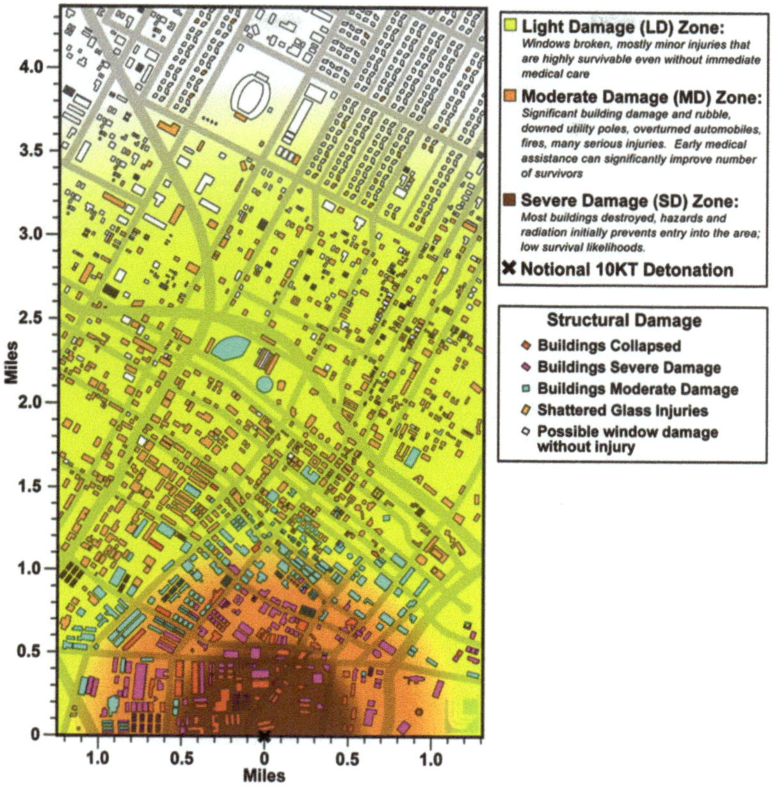

Fig. 1.1 Damage zones resulting from a domestic nuclear detonation (Figure Credit: Lawrence Livermore National Laboratory)

In addition to ionizing and thermal radiation, the detonation creates a brilliant flash of light that can cause temporary blindness to those outdoors over 5 miles away. This effect could go further if there is good visibility or clouds to reflect the light, or if the event occurs at night. "Flash blindness" can even occur if the victim is not looking in the direction of the detonation. It can last several seconds to minutes. Although this effect does not cause permanent damage, the sudden loss of vision to drivers and pilots could cause a large number of traffic casualties and make many roads impassable.

Another long-range prompt effect, which is poorly understood, is glass breakage. Most of the injuries outside of the Murrah building in the 1995 Oklahoma City bombing were caused by this phenomenon [13]. Extrapolating from more recent work on conventional explosives [14], a 10 kt explosion could break certain types of windows (e.g., large monolithic annealed) over 8 miles away. Also noted in this same study was the tendency for glass to fail catastrophically even at extreme ranges, causing severe injury to those behind it. NATO medical-response planning documents [11] for nuclear detonations state that "... missile injuries will predominate.

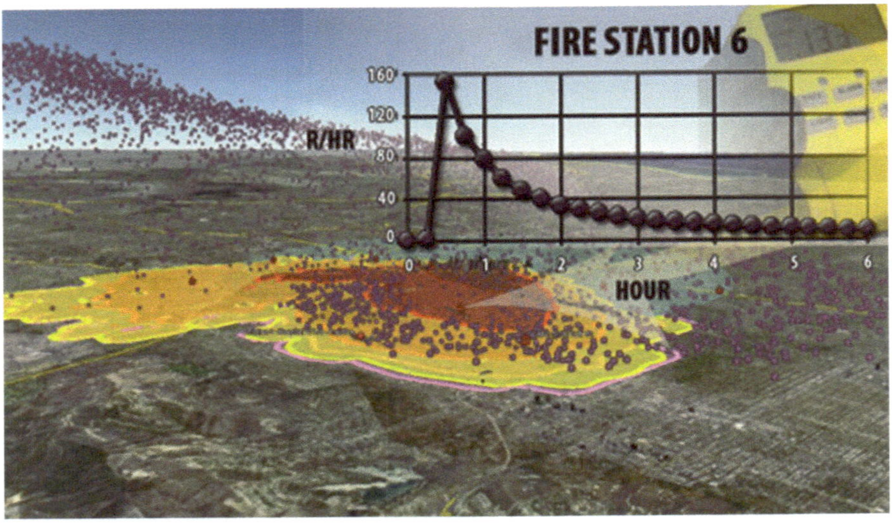

Fig. 1.2 Example of fallout pattern from a 10 kt ground-level detonation (Figure Credit: Lawrence Livermore National Laboratory)

About half of the patients seen will have wounds of their extremities. The thorax, abdomen, and head will be involved about equally." A significant number of victims from Nagasaki arriving at field hospitals exhibited glass-breakage injuries.

The primary delayed effect from a ground-level nuclear detonation is from 'fallout' (Fig. 1.2). Fallout is generated when the dust and debris excavated by the explosion combine with radioactive fission products and are drawn upward by the heat of the event. This cloud rapidly climbs through the atmosphere, up to 5 miles high for a 10 kt, and highly radioactive particles coalesce and drop back down to earth as it cools. It is important to note that Hiroshima and Nagasaki did not have significant fallout because their detonations occurred at altitude.

The hazard from fallout comes not from breathing the particles, but from being exposed to the ionizing radiation they give off after they have settled on the ground and building roofs. Radiation levels from these particles will drop off quickly: most (55 %) of the potential exposure occurs in the first hour, and 80 % occurs within the first day. Although they are highly dependent on weather conditions, the most dangerous concentrations of fallout particles (i.e., potentially fatal to those outside) occur within 10 miles downwind of the event and are clearly visible as they fall, often the size of fine sand or table salt [15].

1.5 Recent Research

The results of recent modeling [16] indicate that a modern urban environment can greatly mitigate some of the effects of a low-yield nuclear detonation. For example, thermal burns from the heat of the initial explosion, primarily a line-of-sight

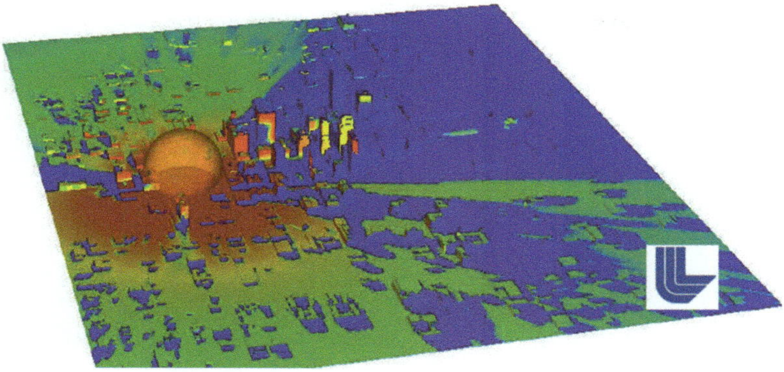

Fig. 1.3 Integrated thermal flux from a 10 kt ground-level nuclear detonation in a small U.S. city (Figure Credit: Lawrence Livermore National Laboratory)

phenomenon, can be greatly reduced in an urban environment where structures can block the thermal radiation. Figure 1.3 shows how building shadows can protect the outdoor population from significant thermal exposure by modeling conducted at Lawrence Livermore National Laboratory [17].

Detailed MCNP Models developed at Applied Research Associates (ARA) [18] and Los Alamos National Laboratory have shown similar reductions in injuries from the initial radiation produced in the first minute of a nuclear explosion. Figure 1.4 demonstrates the nonsymmetrical reduction in radiation exposure by the urban environment. The right side of the image represents an unobstructed exposure from a 10 kt surface detonation as compared to the reduction of outdoor radiation levels indicated in the left side of the image. Like the thermal analysis, these studies indicate that the ambient, outdoor radiation levels from a low-yield, ground-level nuclear detonation in an urban environment could be significantly reduced.

Unlike prompt effects, which occur too rapidly to avoid, health effects from fallout can be mitigated by leaving the area before the fallout arrives or by taking shelter from it. Although some fraction of ionizing radiation can penetrate buildings, shielding offered by walls and distance from outdoor fallout particles can easily reduce exposures by a factor of 10 or more, even in common urban buildings.

The quality of shelter is defined by a protection factor (PF), which is equal to the ratio of outside dose rate divided by inside dose rate. Like sunscreen's SPF, the higher the PF value, the lower the exposure compared to the exposure of an unsheltered person in the same area. Figure 1.5 above shows sample PF estimates based on evaluations conducted in 1973 [19] for typical structures during that era.

Efforts are under way to update the analysis of the level of protection that modern buildings could provide from fallout radiation. Figure 1.6 shows an analysis of more detailed urban structures completed at Lawrence Livermore National Laboratory, in which most areas (shown in green or blue) had PFs greater than 10; which is considered adequate by the Federal Planning Guidance.

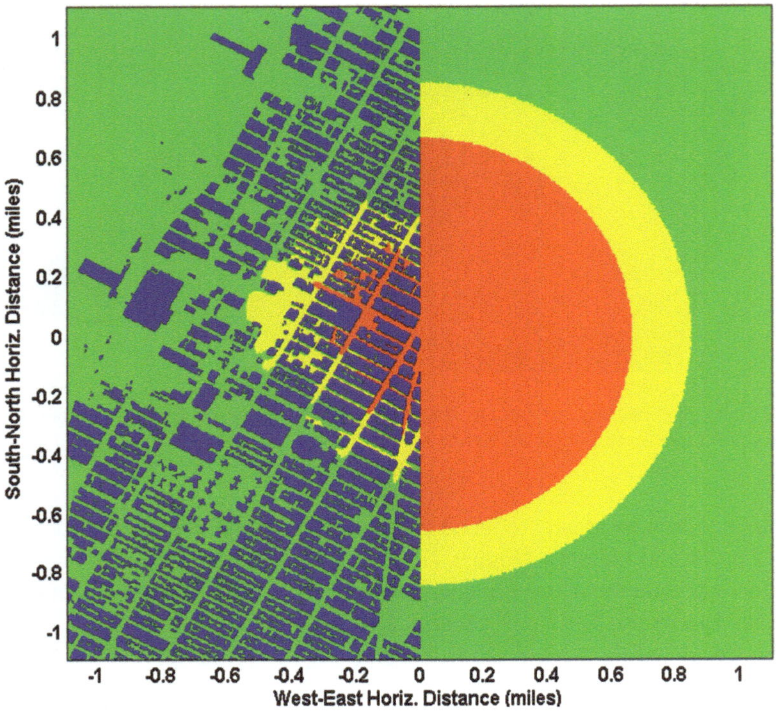

Fig. 1.4 Outdoor casualty areas for high rise urban area (*left*) and for an open field (*right*) from a 10-kt IND; *red* >800 rads (lethal), *yellow* 100–800 rads (injurious to lethal), *green* < 100 rad (non injurious) (Figure Credit: Applied Research Associated, LLC)

Scientists at Lawrence Livermore National Laboratory are developing the capability to assess the quality of urban shelter with respect to nuclear fallout on a regional level [20]. With the results of this assessment (see Fig. 1.7), planners and responders can estimate: (a) the protection provided by existing buildings to fallout radiation, (b) the effectiveness of shelter strategies using existing buildings, and (c) approximate radiation exposures if these shelter strategies were to be used.

Other effects, such as the electromagnetic pulse (EMP) and fires, also need to be considered in response planning and are areas of ongoing research. For a ground-level detonation, most EMP effects will be limited to the blast-damage zones, with a few, random, longer-range disruptions occurring a few miles beyond. Although the possibility of a 'firestorm' is unlikely given modern construction, there will be a large number of small, disparate fires started by thermal and blast effects (generally around the 1-mile perimeter), which could spread and coalesce if not mitigated [24].

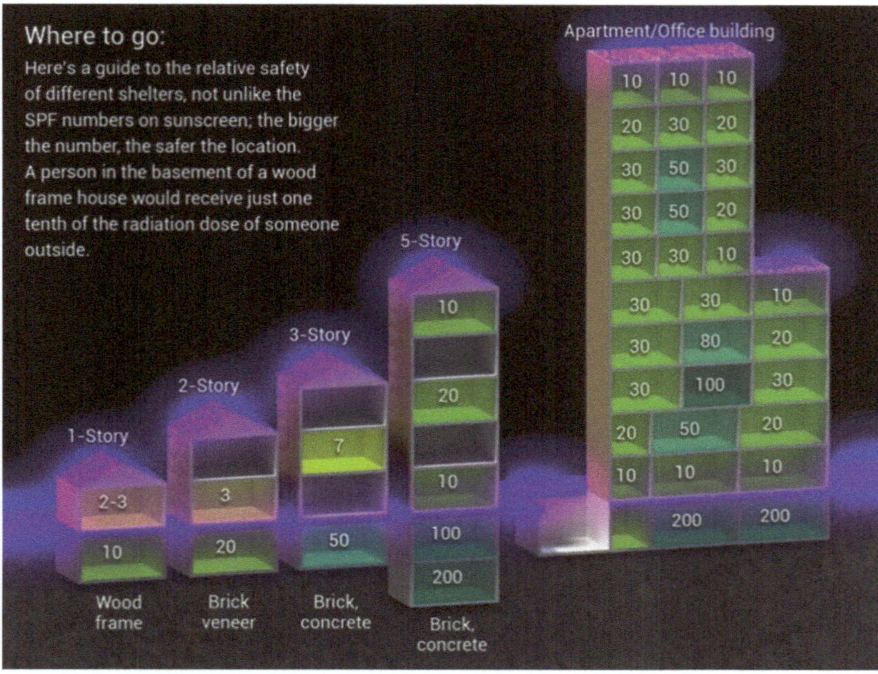

Fig. 1.5 PF by building and by location within building (Figure Credit: Lawrence Livermore National Laboratory)

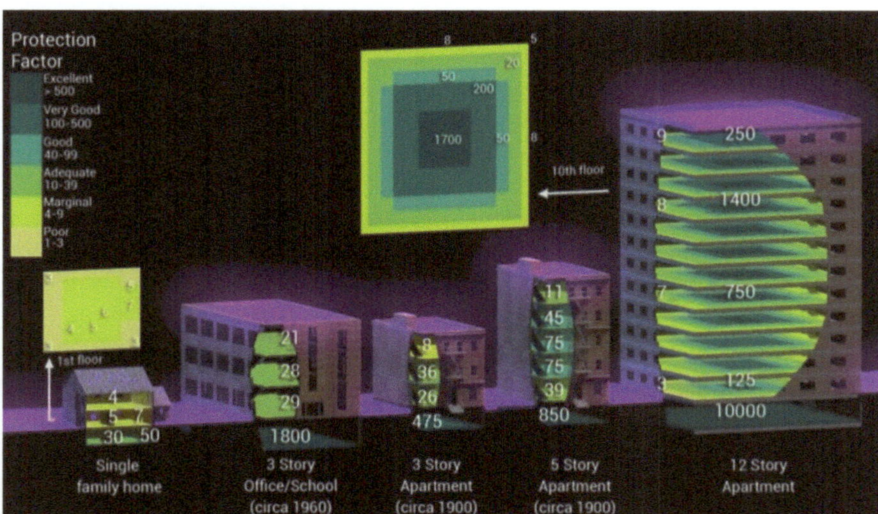

Fig. 1.6 Protection provided by typical urban buildings (Figure Credit: Lawrence Livermore National Laboratory)

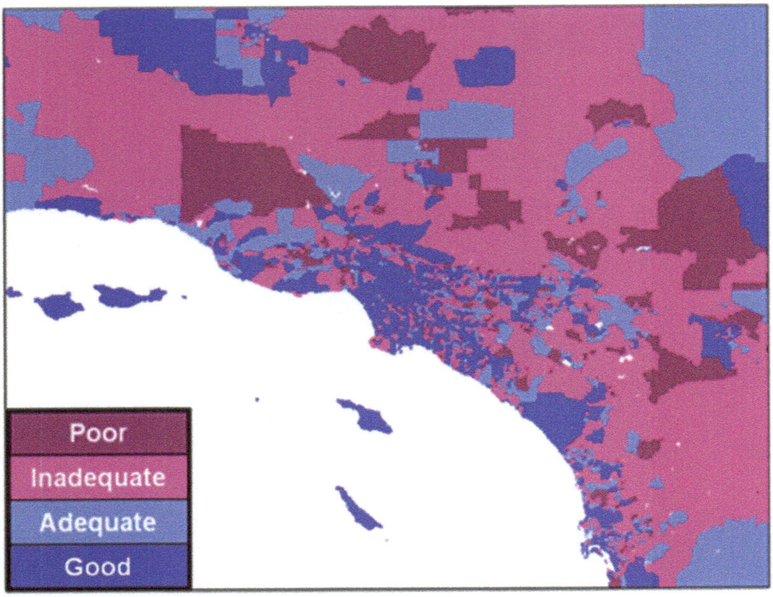

Fig. 1.7 Illustrative evaluation of the protection offered by local (nearby) shelter in the Los Angeles Basin (Figure Credit: Lawrence Livermore National Laboratory)

1.6 Key Response Planning Factors

As stated in the outset of this paper, the end goal of this activity is to build the scientific foundation for responding to large-scale disasters utilizing a nuclear detonation as a case study. Planning Guidance for Response to a Nuclear Detonation, 2nd Edition, produced by a Federal interagency committee led by the Executive Office of the President, Office of Science and Technology Policy [16], is the result of a collaborative effort across many federal departments and agencies. It utilized some of the latest research discussed in this paper and identifies key recommendations in order to respond to and recover from an IND incident. The document identified a zoned approach to facilitate response planning, with the key zones defined as:

- Light Damage Zone: Windows mostly broken; injuries requiring self- or outpatient-care.
- Moderate Damage Zone: Significant building damage and rubble, downed utility poles, overturned automobiles, fires, many serious injuries; greatest life-saving opportunities.
- Severe Damage Zone: Most buildings destroyed; radiation prevents entry into the area; lifesaving not likely.
- Dangerous Fallout Zone: Area where large doses could be delivered to the unsheltered public and emergency responders in a short period of time. This is the dark purple area in Fig. 1.8.

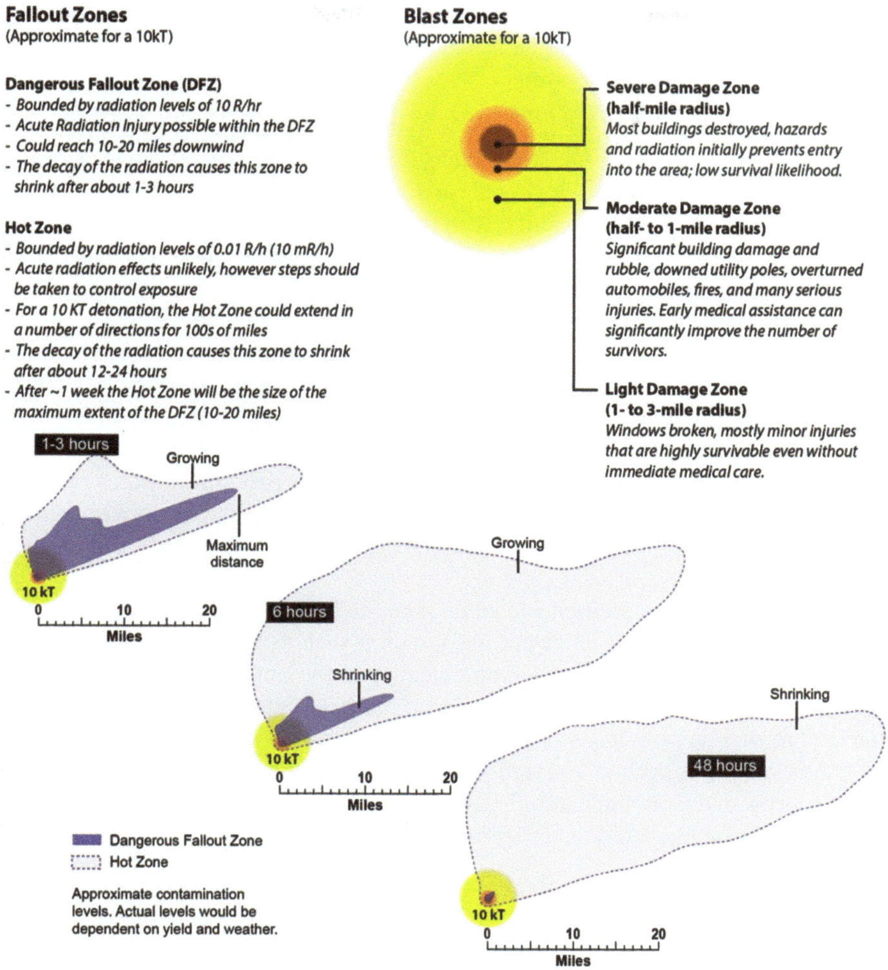

Fig. 1.8 Description of key response planning zones (Figure Credit: Lawrence Livermore National Laboratory)

- Hot Zone or 0.01 R/h Boundary: Areas where emergency operations can be safely performed provided that responders take appropriate planning and dose monitoring and control measures. This is the light purple area in Fig. 1.8.

The Key Response Planning Factors for the Aftermath of Nuclear Terrorism report [21] provided the scientific basis necessary to identify the following response objectives:

- Seek adequate shelter—the most critical lifesaving action for the public and responders is to seek adequate shelter (PF of 10 or more) for at least the first hour.

- Protect response personnel—initial responder efforts should be spent on making high-range dose-rate measurements within their shelter.
- Support regional situation assessment.
- Develop an informed evacuation strategy.

The National Council of Radiation Protection and Measurement Report Number 165; Responding to Radiological and Nuclear Terrorism: A Guide for Decision Makers [22] provides additional scientific backing for the response strategies discussed above. This report is a comprehensive analysis of key decision points and information needed by decision makers at the local, regional, state, and federal levels to help them respond to radiological or nuclear terrorism incidents.

DHS has continued to develop and use the supporting science for response planning activities. Several FEMA regions have developed IND regional joint planning guides and response playbooks to supplement their All Hazards Plans using community specific impact assessments. With every community assessment, new discoveries are made to inform and improve the planning process. FEMA continues to work with responders, emergency managers, and regional planners to better understand how to apply the supporting science and identify community needs.

FEMA continues to facilitate federal, state, and local agency working groups that develop improved guidance and information such as the Improvised Nuclear Device Response and Recovery; Communicating in the Immediate Aftermath [23]. The updated analysis and resulting guidance has been integrated into the training curriculum of FEMA's Incident Management Assistance Teams and mainstream education programs such as Harvard's "Radiological Emergency Planning: Terrorism, Security, and Communication" [24] and Georgetown University's Emergency and Disaster Management Master's Program [25].

1.7 Conclusion

Recent advances in analyzing the effects of a nuclear detonation in an urban area have addressed a number of difficult issues and greatly improved our ability to reduce the consequences of such a horrific event. However, considerable research challenges remain. It is important to note that although sound science is the cornerstone of good response planning, it must be tempered with the unique issues, operational realities, and constraints of emergency-response capabilities in each community. As such, each community may reasonably adopt different response strategies based on the same technical analysis.

This is the cornerstone of the Disaster Management Analytic Framework (DMAF) which provides a flexible process for (1) developing and using sound science to (2) inform community specific assessments that will (3) support the development of guidance that (4) drives regional and national preparedness. The DMAF can be applied to a variety of natural and manmade catastrophic events involving large-scale incident response, and is especially useful for technical hazards that often confound traditional response planning efforts. DMAF facilitates scientific consensus and the development of comprehensive planning guidance in support of emergency managers and response operations.

Acknowledgments Lawrence Livermore National Laboratory would like to acknowledge the leadership and expertise of the Department of Homeland Security's Science and Technology, Domestic Nuclear Detection Office, and the Federal Emergency Management Agency. The authors gratefully acknowledge the insights and support of the Modeling and Analysis Coordination Working Group, a technical working group collaborating on key aspects of nuclear effects modeling whose member include the following contributors: Michael Dillon, John Nasstrom, Kevin Foster, Steve Homann and Gayle Sugiyama at Lawrence Livermore National Laboratory; Steve Maheras at the Pacific Northwest National Laboratory; Kyle Millage and Joe Madrigal at Applied Research Associates; and Jeffery Johnson at Oak Ridge National Laboratory.

LLNL-TR-490770. This work performed under the auspices of the U.S. Department of Energy by Lawrence Livermore National Laboratory under Contract DE-AC52-07NA27344. The Department of Homeland Security sponsored part of the production of this material under the same contract.

References

1. Federal Emergency Management Agency (2011) A whole community approach to emergency management: principles, themes, and pathways for action. Federal Emergency Management Agency, FDOC 104-008-1, December 2011. US Federal Emergency Management Agency, Washington, DC. http://www.fema.gov/media-library-data/20130726-1813-25045-0649/whole_community_dec2011__2_.pdf
2. Buddemeier BR, Suski N (2011) Improvised nuclear device case study, an analytic framework for disaster management. Lawrence Livermore National Laboratory, LLNL-TR-490770, July 2011. Lawrence Livermore National Laboratory, Livermore
3. House of Representatives (2007) Making emergency supplemental appropriations for the fiscal year ending September 30, 2007, and for other purposes. Report 110–107 ed 2007. U.S. Government Printing Office, Washington, DC. http://www.gpo.gov/fdsys/pkg/CRPT-110hrpt107/html/CRPT-110hrpt107.htm
4. Department of Homeland Security (2008) Planning guidance for protection and recovery following radiological dispersal device (RDD) and improvised nuclear device (IND) incidents, 73 ed Federal Register, pp 45029–45049. The federal register is a US Government Publication. http://www.gpo.gov/fdsys/pkg/FR-2008-12-01/pdf/E8-28175.pdf

5. Homeland Security Council Interagency Policy Coordination subcommittee for Preparedness & Response to Radiological and Nuclear Threats (2009) Planning guidance for response to a nuclear detonation. Office of Science and Technology Policy, Executive Office of the President, January 2009. http://www.epa.gov/radiation/docs/er/planning-guidance-for-response-to-nuclear-detonation-2-edition-final.pdf
6. Environmental Protection Agency (2013) PAG manual; protective action guides and planning guidance for radiological incidents, March 2013. Environmental Protection Agency, Washington, DC. http://www.epa.gov/radiation/docs/er/pag-manual-interim-public-comment-4-2-2013.pdf
7. National Academy of Sciences (2005) Nuclear attack, factsheet created for news and terrorism: communicating in a crisis. National Academy of Sciences, Washington, DC. https://www.dhs.gov/xlibrary/assets/prep_nuclear_fact_sheet.pdf
8. Federation of American Scientists (2014) Analysis of Ready.gov. The was an online article http://fas.org/reallyready/analysis.html http://fas.org/reallyready/Templates/pdf/full_analysis.pdf
9. Davis L, LaTourrette T, Howell DR, Dais L, Mosher D (2003) Individual preparedness and response to chemical, radiological, nuclear, and biological terrorist attacks. RAND Corporation. http://www.rand.org/pubs/monograph_reports/MR1731.html
10. Glasstone S, Dolan PJ (1977) The effects of nuclear weapons, 3rd edn. US Department of Defense and the Energy Research and Development Administration, Washington, DC
11. NATO (North Atlantic Treaty Organization) (2005) NATO handbook on the medical aspects of NBC defensive operations (Nuclear) AMedP-6(C). North Atlantic Treaty Organization
12. Mlaker P, Corley W, Sozen M, Thomton C (1998) The Oklahoma City bombing: analysis of blast damage to the Murrah building. Journal of Performance of Constructed Facilities 12(3):113–119
13. Safety Solutions (2007) Preventing glass from becoming a lethal weapon. Case study 30: preventing glass from becoming a lethal weapon. Safety Solutions Online. Archived from the original on February 13, 2007
14. I. Applied Research Associates (2014) Injury based glass hazard assessment: range-to-effect curves,DACA45-02-D-0004. http://www.dtic.mil/dtic/tr/fulltext/u2/a548169.pdf
15. NCRP (National Council on Radiation Protection and Measurements) (1982) The control of exposure of the public to ionizing radiation in the event of accident or attack. National Council on Radiation Protection and Measurements, Bethesda, MD
16. Buddemeier BR (2010) Reducing the consequences of a nuclear detonation: recent research. Br J. National Academy of Engineering, Washington, DC. https://www.nae.edu/Publications/Bridge/19804/19920.aspx
17. Marrs R, Moss W, Whitlock B (2007) Thermal radiation from nuclear detonations in urban environments. Lawrence Livermore National Laboratory, UCRL-TR-231593, June 2007. Lawrence Livermore National Laboratory, Livermore, CA
18. Kramer K, Li A, Millage K, Madrigal J, Sanchez B (2013) Monte Carlo modeling of the initial radiation emitted by a nuclear device in the national capital region. Defense Threat Reduction Agency, DTRA-TR-13-045, July 2013. https://www.hsdl.org/?view&did=756467
19. D. O. D. DEFENSE CIVIL PREPAREDNESS AGENCY (1973) DCPA attack environment manual. Chapter 6, What planners need to know about fallout, CPG 2-1A6, June 1973. US Federal Emergency Management Agency, Washington, DC. https://training.fema.gov/hiedu/docs/historicalinterest/dcpa%20-%20june%201973%20-%20dcpa%20attack%20environment%20manual%20-%20chapter%206.pdf
20. Nasstrom JS, Foster K, GoldStein P, Dillon MB, Wimer N, Homann S, Sugiyama G (2011) Advances in modeling radiation dispersal device and nuclear detonation effects, ANS EPRRSD. Am Nucl Soc. LLNL-CONF-486512, Lawrence Livermore National Laboratory, Livermore, CA
21. Buddemeier BR, Dillon MB (2009) Key response planning factors for the aftermath of nuclear terrorism. Lawrence Livermore National Laboratory, Livermore, California, LLNL-TR-410067, August 2009. Lawrence Livermore National Laboratory, Livermore, CA

22. National Council on Radiation Protection and Measurements (2010) Responding to a radiological or nuclear terrorism incident: a guide for decision makers, p 165, January 2010. National Council on Radiation Protection and Measurements, Bethesda, MD
23. Federal Emergency Management Agency (FEMA) (2013) Improvised nuclear device response and recovery communicating in the immediate aftermath, June 2013. Federal Emergency Management Agency, Washington, DC
24. S. o. P. H. Harvard (2014) Radiological emergency planning: terrorism, security, and communication. https://ccpe.sph.harvard.edu/programs.cfm?CSID=REP0715
25. S. o. C. S. Georgetown University (2014) Executive master of professional studies in emergency and disaster management. http://scs.georgetown.edu/departments/36/executive-master-of-professional-studies-in-emergency-and-disaster-management/

Chapter 2
Response Tactics for the First 100 Minutes After the Outdoor Detonation of an Explosive Radiological Dispersal Device

Stephen V. Musolino, F. Harper, and J. Pfeifer

Abstract The Department of Homeland Security Science and Technology Directorate sponsored a project to operationalize the science-based guidance for dealing with the consequences of a radiological dispersal device, as published in the scientific literature and adopted by the National Council on Radiation Protection and Measurements. The project will morph this scientific guidance into actionable tools for the first-responder agencies that are designed to assist them in developing a simple, concise, and practical radiological- response plan. This effort also will involve a partnership with four cities to pilot and improve these preparedness materials. The principal goal of this project is to leverage scientific guidance to increase the capability of local agencies to respond to a complex radiological event, and assure an effective, coordinated response in the first 100 minutes after the incident.

2.1 Introduction

A summary has been published of the scientific findings at Sandia National Laboratories derived from over 25 years of experiments on the aerosolization of radioactive materials [1]. Based on these data, guidance intended for first responders, planners, and senior decision-makers was developed to promote effective, science-based, response plans [2, 3]. Subsequently, the original guidance was updated. It now reflects new findings from additional experiments with fewer

S.V. Musolino (✉)
Brookhaven National Laboratory, Upton, NY 11973, USA
e-mail: musolino@bnl.gov

F. Harper
Sandia National Laboratories, Albuquerque, NM, USA

J. Pfeifer
New York City Fire Department, New York, NY, USA

S. Apikyan, D. Diamond (eds.), *Nuclear Terrorism and National Preparedness*,
NATO Science for Peace and Security Series B: Physics and Biophysics,
DOI 10.1007/978-94-017-9891-4_2

uncertainties on the potential ballistic fate of the radioactive material, so offering new advice on the tactics of the response [4].

2.2 RDD Planning Guidance Development and Piloting

Recently, the Department of Homeland Security Science and Technology Directorate initiated a project to promote the use of this scientific information to guide the planning efforts of the local radiological dispersal device (RDD) response teams. Hence, this effort will develop simple, concise, and practical tactics and tools to support their planning processes and improve their capability to initiate an effective response in the first 100 minutes of a radiological crisis. The goal of this project is to develop actionable guidance and planning tools for first responders to pilot the materials in four cities. The feedback and lessons learned will be incorporated into the final deliverables.

The first 100 minutes is a critical time in an RDD response because the local response agencies will be challenged to make critical decisions under highly stressful, chaotic conditions. If these choices and tactics are not optimal in resolving the crisis in the first 100 minutes, it is likely the emergency phase of the response will persist for a much longer, meaning that local officials may lose control of the situation and jeopardize their credibility with the public. This unacceptable outcome can be avoided. With proper tactical planning based on scientific guidance, a prompt, effective response can be mounted in the first 100 minutes, so that the responders effectively carry out the correct actions to save lives and protect the responders and the general public, and stabilize the aftermath of the incident. The goal of the project is to increase the capability of local responders to successfully manage the complexities posed by the consequences of an RDD.

2.3 Tactics for the First 100 Minutes

There is extensive expertise and hardware within the responder community to deal with an uncontrolled release of hazardous material. HAZMAT teams already are equipped and trained to work in an all-hazards environment, including one with ionizing radiation. Thus, the aftermath of an RDD generally can be viewed as similar to an improvised explosive device, coupled with large spill of radioactive material, giving rise to some unique complications that planning and preparedness can mitigate. Similar to the recent emergence of the hazard from the Ebola disease, the HAZMAT team has the skills and resources needed, but some additional planning at the tactical level is required to prepare for the response to specific conditions and challenges. Hence, the response agencies need only to adapt and coordinate their existing capabilities, tactics, and resources to deal with the complexities posed by the consequences of the explosion of an RDD.

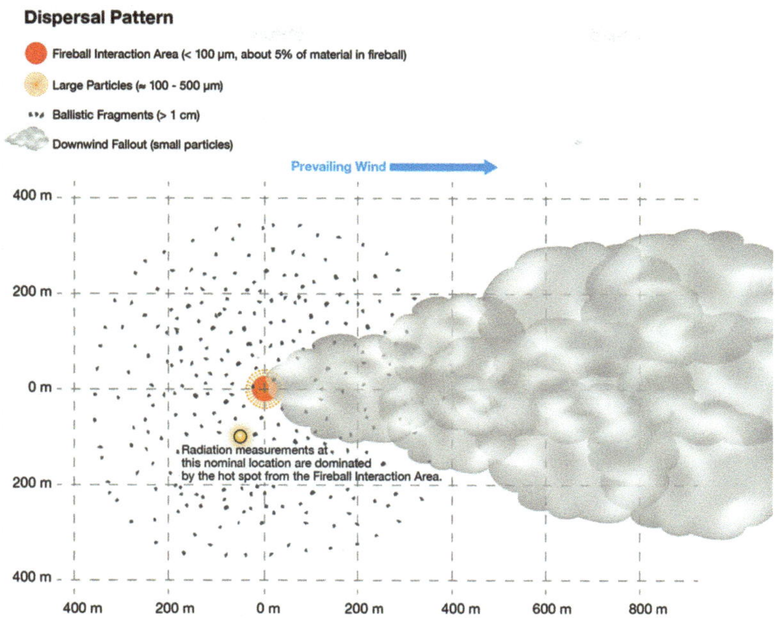

Fig. 2.1 Notional representation of dispersal possibilities. One or more of these fates are possible depending on the form of the radioactive material and the design of the device. Note: These patterns are representative only of the behavior of the radioactive material in the device. Chunks of non-radioactive debris could fall beyond the indicated range for ballistic fragments [4] (The figure was reprinted with permission from the Health Physics Journal and the Health Physics Society)

Because a terrorist incident is expected to occur without warning, early protective actions and decisions must be made in advance so that the preplanned actions become instinctive ones. In principle, this is because the design of the device, and the amount of radioactive material it disperses will be unknown at time of detonation. Thus, it will not be readily apparent whether the contamination is highly localized at the point of detonation, widespread from a plume of aerosol that can deposit contamination over a long distance downwind, or something in-between. Figure 2.1 illustrates the range of possibilities for the fate of the radioactive material.

In planning for the first 100 minutes of an RDD response, first responders need not reinvent their procedures for incident command, medical triage, or emergency messaging, but rather coordinate their actions with other response agencies and incorporate RDD-specific tactics into their concept of operations:

Tactic-1

Initially, the local responders will know only that an explosion has occurred with an associated radiological signature, possibly from one field measurement near the point of detonation. They will not know any of its characteristics, such as those illustrated in Fig. 2.1. Lacking more information, sheltering is recommended to protect

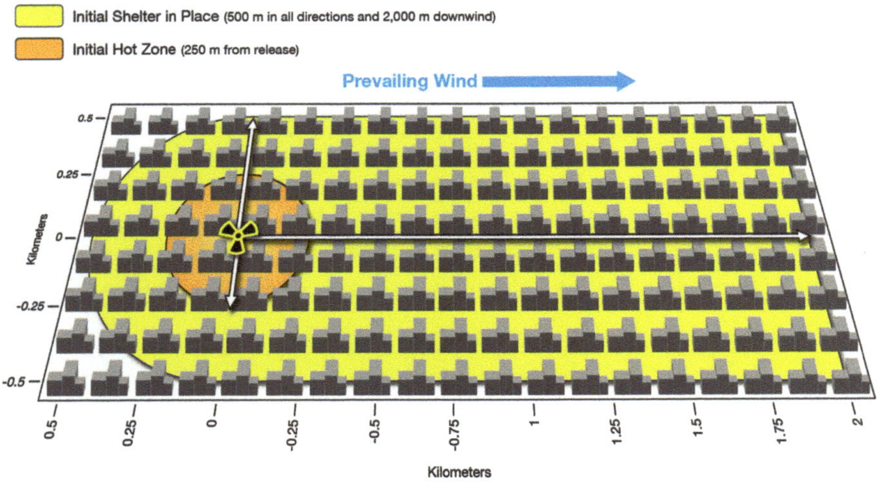

Fig. 2.2 Recommended boundaries for initial Shelter-in-Place and Hot Zones [4] (The figure was reprinted with permission from the Health Physics Journal and the Health Physics Society)

the public until the actual contamination "footprint" is measured and mapped. In this initial period, two tactical default hazard-boundaries are recommended:

1. Define the initial Hot Zone as a 250 m radius around the point of detonation. Do not decide anything based on the perceived direction of the wind, especially in an urban setting where the wind field can be very complex [2, 4]. Later, after collecting field measurements, redefine the Hot Zone based on actual contamination levels and the recommendation of the National Council on Radiation Protection and Measurements of 0.1 mGy h^{-1} [3].
2. Define the boundary to the Shelter-in-Place Zone as 500 m around the point of detonation, and 2,000 m in the direction of the prevailing wind.

Figure 2.2 depicts these recommendations.

Tactic-2

As soon as possible, record the radiation readings near the point of detonation to identify whether or not there is an obvious coherent hotspot. Observe and record the extent to which broken windows are apparent, and the approximate diameter of the crater. A coherent hotspot is an indicator of a large aerosol fraction, and range of broken windows and crater size indicates the amount of explosive (large or small).

Tactic-3

At approximately 1 km, transect (cut across) the Shelter-in-Place Zone and record radiation measurements along this path. The results will indicate whether or not there is contamination over a long distance, and where to conduct the next survey. If null results were obtained, move the next transect closer to the point of detonation, and visa versa if contamination is present.

Tactic-4

Based on the existing inventory of radiation detectors within the city's responder community, and a predetermined plan, rapidly collect and map measurements at long distances (2–10 km) and at 360° around the place of detonation.

Assessment

Consolidate the radiological data from Tactics 2, 3, and 4 on to a map to visualize the information and identify the largely uncontaminated areas surrounding the affected area. Share this information with the public to calm the many concerned citizens, and then identify where others should remain sheltered until directed to evacuate along uncontaminated routes.

2.4 Conclusions

The tactics described above serve as the starting point for a more fully developed tactical RDD response guidance and preparedness materials. The good news is that we know enough from research about RDD detonations to assist first responders in correctly scripting their decision points, and response tactics in the first 100 minutes, allowing cities to prepare through written response plans and exercises. Through the development and piloting of RDD preparedness materials over the next 2 years, expectedly cities will tailor the concepts outlined here, improve them, and possibly develop other new and pragmatic tactics to promote success in the first 100 minutes of an RDD response.

Acknowledgements The Department of Homeland Security Science and Technology Directorate funded the pilot project.

Notice: This manuscript has been authored by Brookhaven Science Associates, LLC under Contract No. DE-AC02-98CH1-886 with the U.S. Department of Energy. The United States Government retains, and the publisher, by accepting the article for publication, acknowledges, a world-wide license to publish or reproduce the published form of this manuscript, or allow others to do so, for the United States Government purposes.

References

1. Harper FT, Musolino SV, Wente WB (2007) Realistic radiological dispersal device hazard boundaries and ramifications for early consequence management decisions. Health Phys 93:1–16
2. Musolino SV, Harper FT (2006) Emergency response guidance for the first 48 hours after the outdoor detonation of an explosive radiological dispersal device. Health Phys 90:377–385
3. National Council on Radiation Protection and Measurements (2010) Responding to a radiological or nuclear terrorism incident: a guide for decision makers, NCRP report 165. National Council on Radiation Protection and Measurements, Bethesda
4. Musolino SV, Harper FT, Buddemeier B, Brown M, Schlueck R (2013) Updated emergency response guidance for the first 48 hours after the outdoor detonation of an explosive radiological dispersal device. Health Phys 105:65–73

Chapter 3
Operational Framework for Recovery from an Attack Involving a Radiological Dispersal Device/Improvised Nuclear Device

Ellen Raber, Robert P. Fischer, and Mark Sutton

Abstract This paper adapts a six-phase U.S. Office of Science and Technology Policy biological and chemical incident preplanning response and recovery framework for a large-scale radiological terrorist attack. This framework depicts preplanned decisions for a radiological emergency event to ensure protection of public health and the environment. Recommendations for operational-level details across the remediation space of characterization, decontamination, and clearance are provided as well as an overview of the current technologies available and the gaps that are important to consider in the timeline for recovery. Examples of the use of this framework applied to radiological preplanning are also discussed.

3.1 Introduction

Recovering from a radiological attack is a complex process requiring the successful resolution of numerous challenges. National policies and regulations address preparedness goals and organizational structure, but they do not tell responders how to perform remediation. This article highlights features of a national-level framework that has been developed to guide a risk-based decision process and inform decision makers of the questions that must be addressed and supported by best practices to optimize recovery of functions at affected facilities or areas. Essential considerations include (1) specifying the emergency-response actions needed at the onset of a radiological incident; (2) determining the extent of contamination and whether an actual or potential impact to health, property, or the environment exists; (3) determining the actions needed to restore essential facilities and/or operations; (4) determining whether remediation is needed and a process to evaluate options; (5) identifying risk-based clearance goals; and, (6) selecting appropriate decontamination technologies to meet those goals. In order to

---

E. Raber (✉) • R.P. Fischer • M. Sutton
Lawrence Livermore National Laboratory, Livermore, CA, USA
e-mail: raber1@llnl.gov

S. Apikyan, D. Diamond (eds.), *Nuclear Terrorism and National Preparedness*, NATO Science for Peace and Security Series B: Physics and Biophysics,
DOI 10.1007/978-94-017-9891-4_3

be successful, it is essential that stakeholders are involved. Unique environmental, economic, and cultural considerations must always be considered during any site- and incident-specific radiological recovery. Additionally, preplanning investments in critical areas will help to shorten the timeline for recovery and allow communities to be better prepared should such an incident occur.

3.2 A Decision Framework for Recovery from a Radiological Attack

A successful recovery from a large-scale radiological incident will involve scenario- and site-specific decisions requiring an integrated systems approach. Numerous documents and policies have been produced at the national level addressing various topics related to preparedness training. The *National Preparedness Guidelines* [1] states the core preparedness goal for the nation. The National Response Framework (NRF) is a "guide to how the nation conducts all-hazards incident response" when managing domestic incidents [2]. The *Comprehensive Preparedness Guide 101, version 2.0*, provides Federal Emergency Management Agency (FEMA) guidance on the fundamentals of planning and developing emergency operations plans [3]. It illustrates the relationships among national policies, regulations, standards, and other initiatives at state and local levels that must be coordinated during remediation. Many other policies and plans are also relevant to consequence management following a radiological attack; however, their high-level content contains little guidance for conducting the numerous operational-level activities associated with remediation. None of the documents, plans, or acts bring all the required operations together in policy-level guidance that tells responders how they should perform remediation and recovery.

To fill the gap, we have reviewed previously developed frameworks for chemical/biological/radiological (CBR) incidents, which can serve as a starting point for approaching a more operational-level decision tool [4, 5]. Additionally, the Department of Homeland Security recently published *Planning Guidance for Protection and Recovery Following Radiological Dispersal Device (RDD) and Improvised Nuclear Device (IND) Incidents*, and *Key Planning Factors for Recovery from a Radiological Terrorism Attack* [6, 7]. These documents use the NRF as an overarching approach to establish preplanning factors that should be considered by decision makers. More recently, the Environmental Protection Agency (EPA) published the *Interim Protective Action Guide and Planning Guidance for Radiological Incidents*, which is very relevant to establishing remediation goals and objectives [8]. The purpose of this paper is to utilize these previous studies and suggest a decision framework for RDD/IND remediation planning and execution that can serve as a model for the nation as a whole in addressing the unique challenges associated with an RDD/IND terrorism attack in an urban area. Our focus is on radiological contamination resulting from the attack rather than on blast, thermal, or other effects, because of the potential for long-term health or contamination issues.

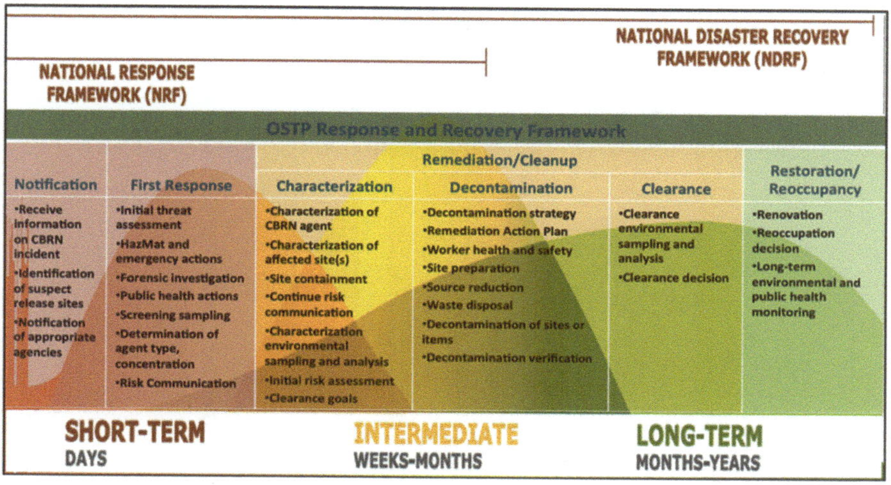

Fig. 3.1 Integration of NRF and OSTP frameworks for key planning factors

In particular, this framework identifies areas necessary for preplanning to guide stakeholder and interagency coordination at federal, regional, state, and local levels, as well as addresses social, economic, and operational interdependencies and limitations that affect remediation and recovery actions. We have utilized and combined the NRF and the *National Disaster Recovery Framework* (NDRF), and incorporated the Office of Science and Technology Policy (OSTP) framework (initially developed for biological incidents), which follows the scheme of six phases for response and recovery arrived at through interagency consensus and approval. Figure 3.1 shows the combined integration of this approach. These six phases are adapted for a radiological incident in a series of detailed decision flowcharts (Figs. 3.2, 3.3, 3.4, 3.5, and 3.6), identifying questions that must be answered from the time that an incident is discovered. Each box within the framework has an associated number (in blue) that is used to link to expanded operational-level guidance and can be referred to in supporting documentation.

There are four overarching concepts that are important to this framework. First, communication to the public must be accurate, timely, and frequent, and the coordination of local, regional, state, and federal public information is critical. Therefore, risk communication is emphasized throughout the phases of response and recovery. Second, although the series of decision flowcharts comprising the framework seem to treat the six phases and steps within them as sequential, many activities will overlap or take place concurrently and some will be recurrent. Third, because the goal is to balance achievable and practical results, an optimization approach is emphasized at each stage of remediation. Optimization is not only important to identify successful cleanup options but also to guide the choice of targets during remediation and restoration and to ensure that limited recovery resources are used effectively. And fourth, our understanding of many technical details important to recovering from a wide-area radiological attack is currently

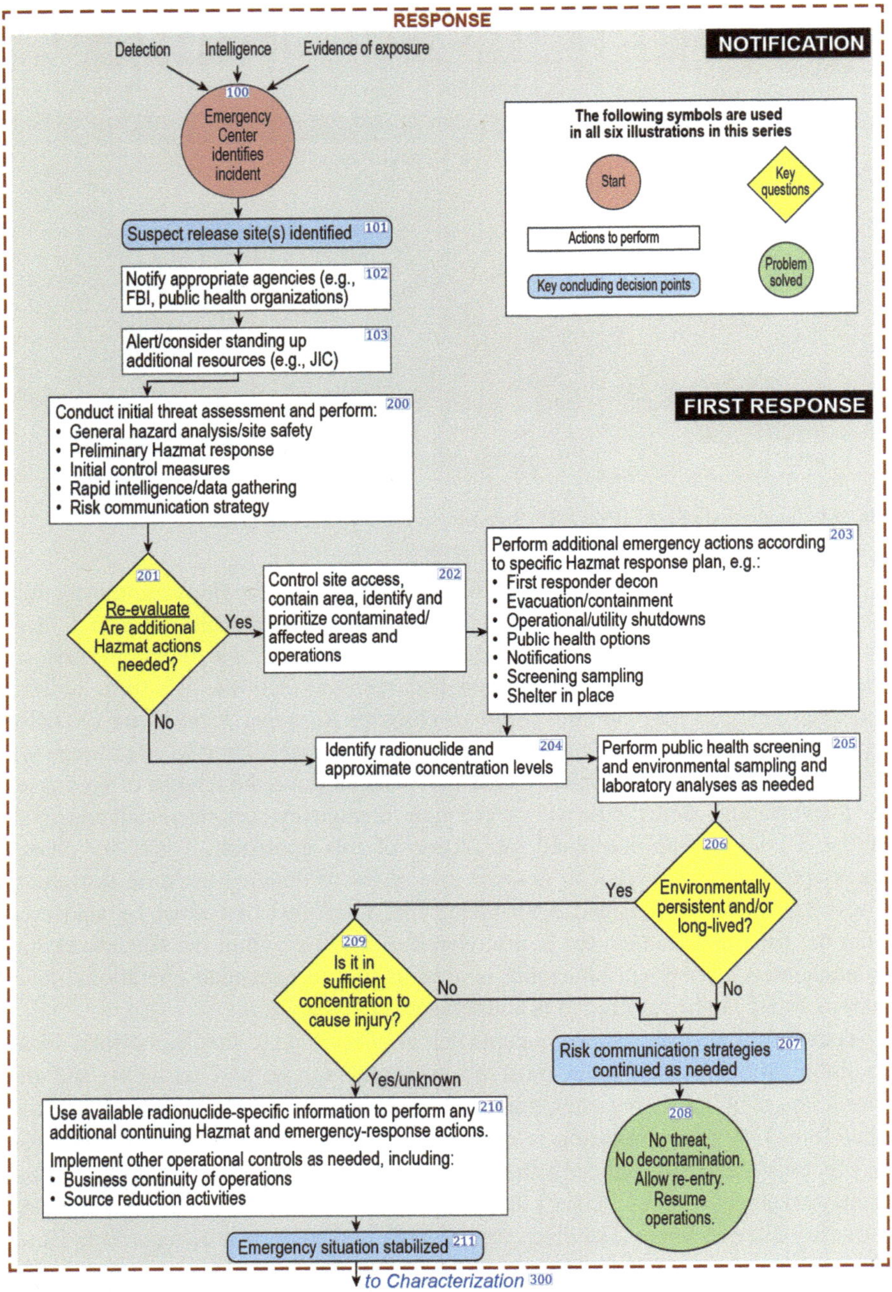

Fig. 3.2 Notification and first response phases

limited and any actionable-level guidance will need to address both current best practices as well as identify gaps in technical capability and resources.

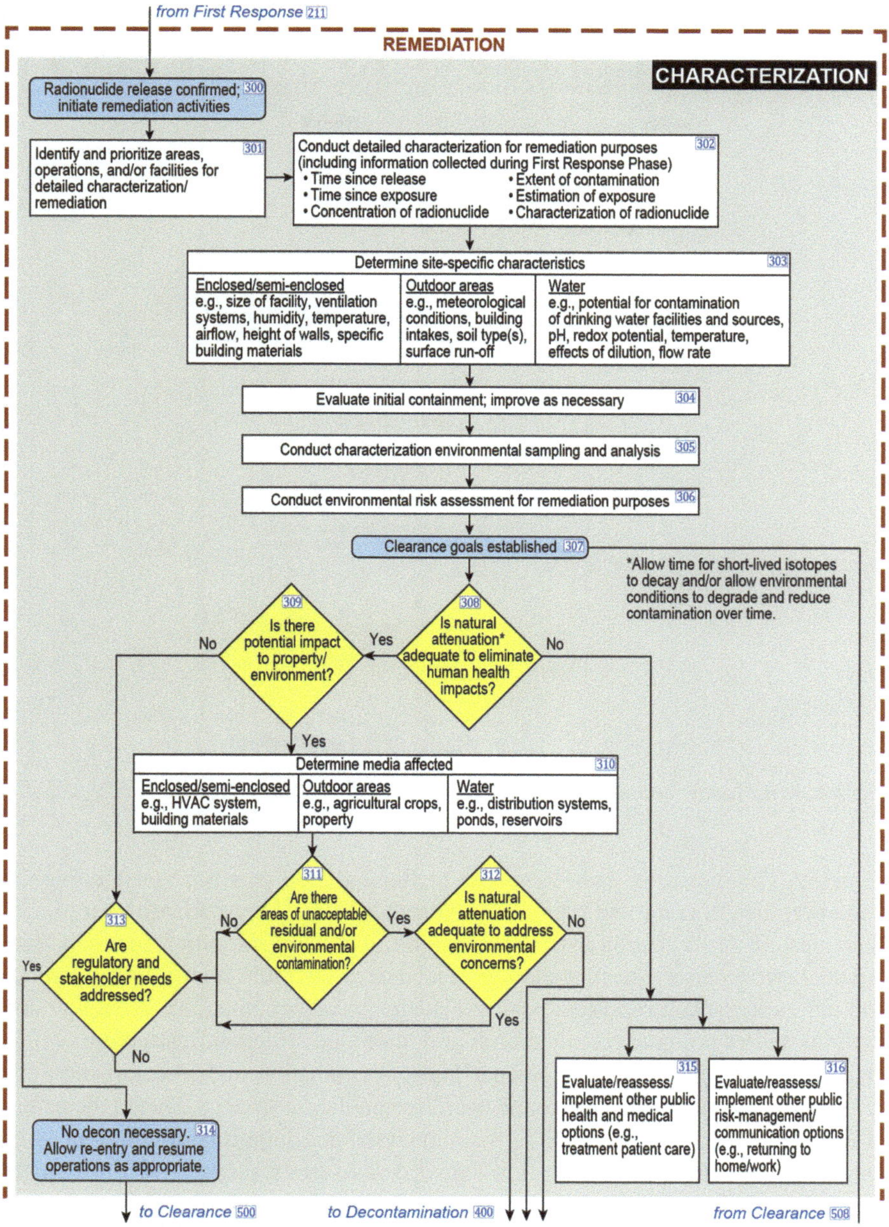

Fig. 3.3 Remediation: characterization phase

Our studies have primarily focused on the Recovery or Remediation Phase consequence management activities (i.e., Characterization, Decontamination, and Clearance) rather than initial phases (Notification and First Response); although the successful transitional phase from Response to Remediation is key for effective

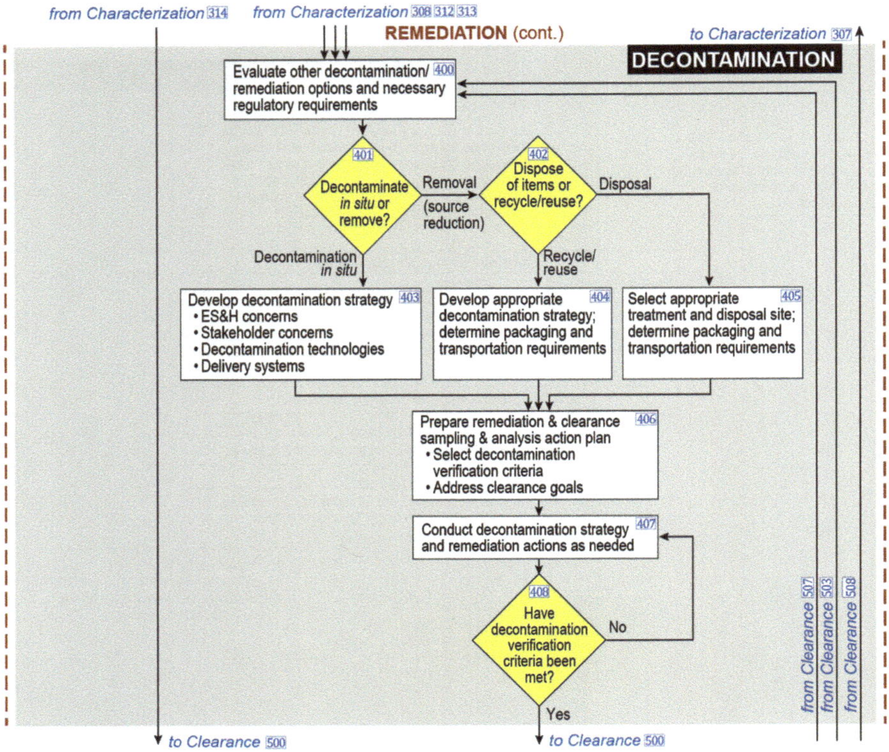

Fig. 3.4 Remediation: decontamination phase

recovery. The transition from Response to Remediation entails understanding the remaining health risks and then addressing those risks by prioritizing assets and functions. In order to optimize the transition, it is very important for individual jurisdictions to prioritize remediation objectives, determine high-priority outdoor and indoor areas that represent health-related/environmental issues, economic disruptions and/or political considerations. Additionally, critical infrastructure (which includes lifeline infrastructure), areas of high contamination, and other jurisdictional priorities, must be integrated into the overall remediation strategy. Due to the nature of radiological contamination, it is recommended that important outdoor areas be remediated or stabilized first to allow safe access to contaminated structures and to prevent the further spread of contamination. Facilities in cleared areas would then follow an additional prioritization for continued remediation until all areas have been cleared for restoration and re-entry.

We are currently evaluating the operational-level information needed with emphasis on the Remediation Phase to allow and support a rapid return to service concept and strategy. This includes a more detailed understanding for Characterization, Decontamination, and Clearance, and the application of that understanding for specific scenarios and infrastructure. This is being accomplished by providing

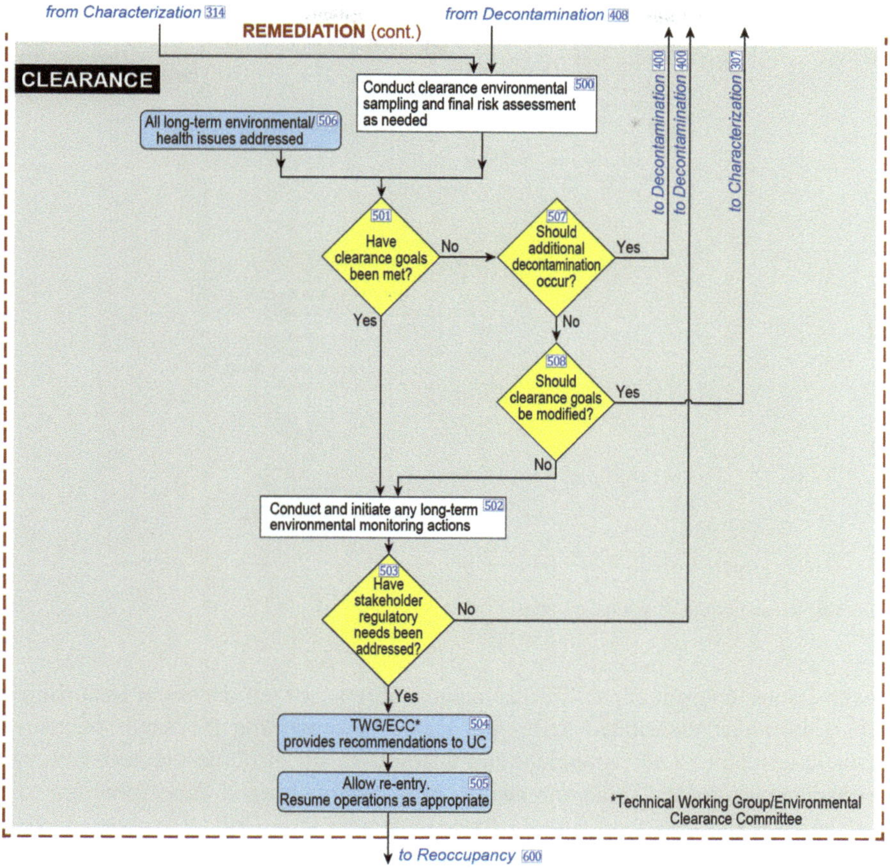

Fig. 3.5 Remediation: clearance phase

more detail and linkages to the existing boxes in the framework as has already been done in detail for chemical and biological incidents [9, 10]. It also involves an understanding of the current/emerging technologies that can be employed and optimized for a radiological event to reduce recovery timelines. Application and choice of technologies will be site- and incident-specific. One important parameter that needs to be considered is the availability of local and national response and recovery capabilities (manpower, equipment). An overview of the three phases for remediation as well as the general state of technology for radiological remediation within those areas is discussed below.

Characterization Characterization provides estimates of how widespread the contamination is by assessing outdoor surface contamination, outdoor air contamination from re-suspension of particles, and indoor contamination of facilities. In general, characterization is fairly robust for radiological analyses. Rapid characterization will quickly determine the extent of contamination and address important public

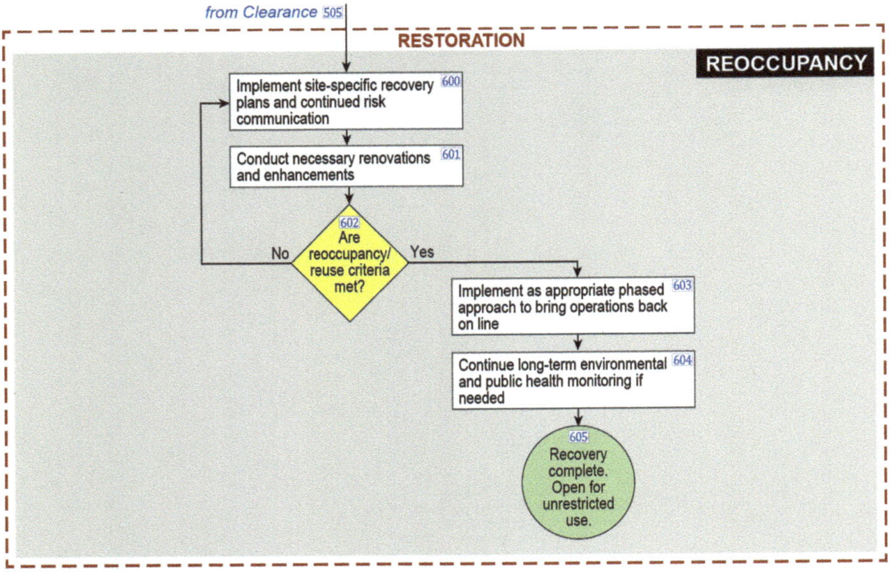

Fig. 3.6 Remediation: reoccupancy phase

health issues. Current capabilities as demonstrated for the incident at Fukushima, Japan show that the United States is capable of supporting thousands of assays to include filters (both particulate and adsorptive) for airborne radioactivity and re-suspension, surface soils, and swipes to determine ground deposition and soil core analysis to understand migration rate impact to dose. Critical analyses needed include gross alpha and beta screening, gamma spectrometry for iodine-131, cesium-134, cesium-137, strontium-89, strontium-90, and others as well as actinide analysis (uranium, neptunium, plutonium, americium, and curium). It is important to note that a multi-faceted modeling and ground-truth approach is necessary to rapidly understand impacts from a radiological incident. This requires a rapid aerial monitoring system as well as in situ and laboratory analysis capabilities. This combined set of tools helps to build public confidence. The remediation timeline could be shortened with more field-based, higher sensitivity prompt analysis methods to include remote and automated analysis systems.

Decontamination The purpose of decontamination is to clean up areas affected by a radiological release that remain a risk to human health, as determined from the Characterization Phase, with the highest priority placed on restoring critical infrastructure to minimize economic and social impacts. Decontamination for the public sector is very demanding and requires rigorous evaluations for several key parameters, including: availability, effectiveness, safety, compatibility, operational requirements, time, waste generation, environmental concerns, and stakeholder needs. Therefore, determining whether decontamination is necessary is a site-specific, incident-specific decision. It is important to determine whether contaminated areas will be treated in situ or whether contaminated materials will be removed and then treated at a different location. Lawrence Livermore National Laboratory (LLNL) has a particular understanding of cesium and how to remediate outdoors in some environments (e.g., carbonate geology). This is based on LLNL's experience in the Marshall Islands, which is discussed in more detail below.

After an incident, the use of stabilization/fixative methods will limit the exposure and further spread of radiological contamination. In particular, fixatives can be applied to indoor and/or outdoor surfaces and can allow agglomeration/stabilization/encapsulation to reduce inhalation health risks. The US EPA is evaluating several approaches, from commonly available fire-fighter materials (e.g., water, foams, retardants) and other locally available materials to rad-specific commercial materials with demonstrated effectiveness, although not as readily available. Once stabilized, an understanding of the fate and transport mechanisms for specific radionuclides is important in determining decontamination strategies and methods. Additionally the decontamination methods will vary and must be evaluated independently for (1) indoor/semi-enclosed environments; (2) outdoor areas (both natural and man-made surfaces; and, (3) water systems (both municipal distribution and natural systems). For example, cesium migration on porous urban surfaces requires immediate action to avoid migration into surface materials. However, short-lived isotopes may not require any decontamination if a combination of stabilization/mitigation methods in conjunction with natural attenuation are found to be sufficient to reduce health risks to an acceptable level. LLNL has developed a more detailed flowchart for evaluating and determining the best level of decontamination for facilities and/or infrastructure. This includes three levels of decontamination depending on surface type and reactivity of the specific radionuclide. Promising wide-scale decontaminants include potential chelation agents, although these are still under development. Lastly, issues with waste generation and treatment need to be addressed and waste minimized and/or stabilized for landfill disposal when possible.

Clearance Clearance is the process of determining whether a specified clearance goal—developed from risk assessment and risk management practices—has been met for a particular contaminant in or on a specific area, site or item. Wide-area clearance will need to be performed in phases, with some parts of an urban area undergoing clearance before others according to prioritization. Understanding and conducting effective risk assessments is key to optimizing risk-management and remediation strategies. Most important is the translation of environmental sampling into expected exposure/activity levels from potential exposure pathways. This allows for the evaluation of risk- and dose-based criteria and translates to probability of disease (e.g., cancer risk) and public health risk. The process of determining "How Clean Is Clean Enough?" and establishing clearance goals involves stakeholder input throughout the decision process and needs to be sensitive to political, social, legal, and cultural factors. Risk-based and dose-based clearance criteria for radiological exposures have been developed by several U.S. agencies [EPA, DOE, NRC, Agency for Toxic Substances and Disease Registry (ATSDR)] as well as the International Commission on Radiological Protection (ICRP). Although these criteria can serve as a starting point, consensus does not exist as to how to apply them in a wide-area event, and this will be an incident- and site-specific decision [11]. Improved fundamental risk-assessment and risk-management approaches, including exposure pathways, are needed.

3.3 Applications and Lessons Learned

The Marshall Islands reclamation project is a real-world example of radiological (cesium-specific) remediation success, but the timelines need to be significantly shortened for a radiological terrorist incident [12]. This previous work started with a detailed wide-area characterization and examined possible decontamination approaches. After initial field and laboratory studies, the approach focused on the displacement of cesium from large areas using potassium (K). Although this was ultimately successful, the method is highly geologic dependent and applications to other sites would need further evaluation. The area has now been repopulated and agriculture has been re-established, although long-term monitoring continues. Details of this project should be reviewed in more depth for application to other potential incidents, including potential applications to Fukushima.

More recently, LLNL has worked with several U.S. government agencies and developed detailed recovery plans for urban facilities and transit systems based on the OSTP adapted framework presented here. For example, we developed detailed rapid return to service guidance based on this framework for the Manhattan Transit Authority to use in the event of a radiological attack. This supporting guidance included (1) monitoring and characterization plans; (2) facility decontamination plans; (3) rolling stock remediation plans; and, (4) waste management plans. The overall decision framework worked well in developing the necessary detailed supporting guidance documents.

3.4 Summary

Restoration and recovery requirements for the civilian sector are very demanding and conflicting for a wide-scale radiological incident and will be incident- and site-specific. Significant regional and national economic drivers will require fast, adequate, and reduced-cost solutions. These will be in conflict with stakeholders who want high assurances that facilities/areas are safe for reoccupancy. Therefore, it is important to optimize these parameters based on an effective operational-level decision framework and to develop supporting preplanning guidance. LLNL has begun work on an operational-level framework based on the OSTP biological framework integrated with the NRF as applied to radiological response and recovery. Successful implementation of the proposed framework is limited by capability gaps (gaps in technical capability and limited resources) underlying the likely requirements facing incident commanders who will manage remediation activities following an RDD/IND incident. Important gaps include (1) staging and availability of local and national response and recovery capabilities (manpower, equipment, and other potential shortages); (2) the need for improved fundamental risk-assessment and risk-management approaches, including the transport and potential spread of radionuclides; (3) the need for improved stabilization and/or decontamination technologies, especially for wide-scale outdoor areas, including how to deal with infiltration or penetration of radionuclides into porous construction materials; (4) lack of supporting information and sampling procedures needed to develop and confirm realistic outdoor clearance goals; and, (5) potentially overwhelming waste-management and disposal issues. Lessons learned from the incident in Fukushima, Japan, and remediation of the Marshall Islands are key to moving forward with a long-term goal of an operational-level framework and supporting guidance for wide-area rapid return to service from an RDD/IND incident.

Disclaimer and Auspices This document was prepared as an account of work sponsored by an agency of the United States government. Neither the United States government nor Lawrence Livermore National Security, LLC, nor any of their employees makes any warranty, expressed or implied, or assumes any legal liability or responsibility for the accuracy, completeness, or usefulness of any information, apparatus, product, or process disclosed, or represents that its use would not infringe privately owned rights. Reference herein to any specific commercial product, process, or service by trade name, trademark, manufacturer, or otherwise does not necessarily constitute or imply its endorsement, recommendation, or favoring by the United States government or Lawrence Livermore National Security, LLC. The views and opinions of authors expressed herein do not necessarily state or reflect those of the United States government or Lawrence Livermore National Security, LLC, and shall not be used for advertising or product endorsement purposes.
This work was performed under the auspices of the U.S. Department of Energy by Lawrence Livermore National Laboratory under Contract DE-AC52-07NA27344.

References

1. Federal Emergency Management Agency (FEMA) (2011) National preparedness guidelines, 1st edn, September 2011
2. FEMA (2011) National disaster recovery framework, September 2011
3. FEMA (2010) Developing and maintaining emergency operations plans, comprehensive preparedness guide 101, version 2.0, November 2010
4. Department of Homeland Security (DHS) and Environmental Protection Agency (EPA) (2009) Draft planning guidance for recovery following biological incidents. https://www.federalregister.gov/articles/2009/08/17/E9-19688/availability-of-draft-planningguidance-for-recovery-following-biological-incidents
5. Raber E, Hirabayashi J, Mancieri S, Jin A, Folks K, Carlsen T, Estacio P (2002) Chemical and biological agent incident response and decision process for civilian and public sector facilities. Risk Anal 22(2):195–202
6. Department of Homeland Security (DHS) (2008) Planning guidance for protection and recovery following radiological dispersal device (RDD) and improvised nuclear device (IND) incidences (Federal Register 73(149):45029–45048), DHS (U.S. Department of Homeland Security), August 1 2008

7. DHS (2012) Key Planning Factors for Recovery from a Radiological Terrorism Incident (authors, Buddemeier BR, Wood-Zika AR, Fischer RP, Valentine JE) LLNL-TR-574123/E1036
8. Environmental Protection Agency (2013) Protective action guides and planning guidance for radiological incidents (Federal Register, Draft for Interim Public Comment), March 2013
9. Raber E, Hibbard JWJ, Greenwalt R (2011) The national framework and consequence management guidance following a biological attack. Biosecur Bioterror: Biodef Strategy Pract Sci 9(3):271–279
10. Lawrence Livermore National Laboratory (2012) Remediation guidance for major airports after a chemical attack, UCRL-TR-562207
11. Raber E, Kirvel R (2008) Health risk assessment for radiological, chemical and biological attacks. Homeland security policy and technology manual. Wiley, December 2008
12. Robison WL, Coronado CL, Bogen KT, Stoker AC (2003) The effective and environmental half-life of Cesium-137 at Coral Islands at the former U.S. nuclear test site. J Environ Radioact 69:207–223

Chapter 4
Responding to Nuclear Terrorism: Is Training of First Responders at Sites of Nuclear Disasters Enhancing Their Preparedness for Incidents Involving Nuclear and Radioactive Material?

Dieter Rothbacher, Oskar Schmidt, and Duncan Brierley

Abstract Preparedness of the first responders' community is largely driven by extensive training programmes, based upon operational procedures and incident awareness and management. Specialised training facilities allow enhancing preparedness for incidents involving radioactive and nuclear material; due to the merging threat posed by radiological, nuclear and explosive (RNE) materials, there has been increased focus on this particular training for first responder and military communities worldwide. Few countries have live-agent training facilities that would even allow the use of open, unsealed radioactive sources.

But there is the need for realistic full force field training to confirm the units' readiness to assume the mission. The accident in the 4th reactor unit of the Chernobyl nuclear power plant that occurred in the early morning of April 26, 1986 is known to have released an enormous quantity of radionuclides from the destroyed reactor. The activity that was released into the atmosphere to generate the fallout that precipitated onto the adjacent territories was approximately 3,330 PBq (90 MCi), not including noble gases. The most affected area, the so-called exclusion zone of Chernobyl, covers an area of approximately 2,600 km^2 (1,000 sq mi) in Ukraine, immediately surrounding the Chernobyl nuclear power plant where radioactive contamination from fallout was highest and public access and inhabitation are still restricted. This exclusion zone is used for radiation protection and response training of first responders. Can this training be done safely and are the benefits outweighing the risks?

D. Rothbacher (✉) • O. Schmidt • D. Brierley
Hotzone Solutions Group, Den Haag, Netherlands
e-mail: dieter.rothbacher@hotzonesolutions.com

S. Apikyan, D. Diamond (eds.), *Nuclear Terrorism and National Preparedness*,
NATO Science for Peace and Security Series B: Physics and Biophysics,
DOI 10.1007/978-94-017-9891-4_4

4.1 Introduction

Preparedness of the first responders' community is largely driven by extensive training programmes, based upon operational procedures, incident awareness and management. Specialised training facilities allow enhancing preparedness for incidents involving radioactive and nuclear material. Due to the emerging multifaceted threats posed by radiological, nuclear and explosive (RNE) materials, increasing focus has been placed on this particular training for first responder and military communities worldwide.

Few countries have live-agent training facilities that would even allow the use of open, unsealed radioactive sources. But there is the need for realistic, full-force field training to confirm the units' readiness to assume the mission.

4.1.1 'Train as You Fight'

The accident in the 4th reactor unit of the Chernobyl nuclear power plant that occurred in the early morning of April 26, 1986 is known to have released an enormous quantity of radionuclides from the destroyed reactor. 3,330 PBq (90 MCi) – not including noble gases – were released into the atmosphere, to generate the fallout that precipitated onto the adjacent territories.

The most affected area – the so-called exclusion zone of Chernobyl – stretches over some 2,600 km^2 (1,000 sq mi) in Ukraine, all around the Chernobyl Nuclear Power Plant. There, the highest radioactive contamination from fallout was recorded, and public access and inhabitation are still restricted.

The exclusion zone is used for radiation protection and response training of first responders.

Can this training be carried out safely and are the benefits outweighing the risks?

This conference paper shall answer those questions.

4.2 The Chernobyl Exclusion Zone

The accident in the 4th reactor unit of the Chernobyl nuclear power plant that occurred in the early morning of April 26, 1986 is known to have released an enormous quantity of radionuclides from the destroyed reactor. At the time of the accident, the 4th reactor unit had been in operation for just over 2 years (since December 1983), and the reactor's core was estimated to have accumulated approximately 2,220 PBq ($6.0 \cdot 10^7$ Ci)[1] of the activity, without taking into account

[1]During the period of the ChNPP accident, units that were not part of the International System of Units (the SI system) were commonly used in the Soviet Union. This article shows the values in the SI system with the original values used by the authors of the references given in brackets where possible.

Table 4.1 Inventory of major long-lived radionuclides in the core of the ChNPP 4th reactor unit [1]

Radionuclide	Mass, kg	Activity	
		Bq	MCi
^{90}Sr	44.24	$2.278 \cdot 10^{17}$	6.158
^{106}Ru	6.935	$8.585 \cdot 10^{I7}$	23.20
^{134}Cs	3.195	$1.531 \cdot 10^{17}$	4.138
^{125}Sb	0.502	$1.920 \cdot 10^{16}$	0.5190
^{137}Cs	81.120	$2.597 \cdot 10^{17}$	7.020
^{144}Ce	33.22	$3.922 \cdot 10^{18}$	106.0
^{236}Pu	$5.814 \cdot 10^{-6}$	$1.I44 \cdot 10^{11}$	0.000003093
^{238}Pu	1.479	$9.377 \cdot 10^{14}$	0.02534
^{239}Pu	412.7	$9.480 \cdot 10^{14}$	0.02562
^{240}Pu	176.0	$1.482 \cdot 10^{15}$	4.005
^{241}Pu	49.11	$1.835 \cdot 10^{17}$	4.958
^{242}Pu	14.16	$2.061 \cdot 10^{12}$	0.0000557

short-lived radionuclides [1]. Including gaseous and short-lived radionuclides, the total activity in the 4th reactor unit at the time of the accident was estimated to be 300,107 PBq ($8{,}111 \cdot 10^6$ Ci) [2]. About 3.5 % of this activity was released into the atmosphere, to generate the fallout that precipitated onto the adjacent territories. According to different estimates, the total radioactive release (not including noble gases) was approximately 3,330 PBq (90 MCi) (Borovoy 1990). The release of the radionuclides from the destroyed reactor unit took a fairly long time, and continued from April 26, 1986 to May 6, 1986 (Table 4.1).

4.3 "Train as You Fight" Systems Approach to Training for First Responders at Sites of Nuclear Disasters

Here is a description of how Hotzone Solutions (HZS) has applied the globally recognised but usually generic 'Systems Approach to Training' concept, used by many organisations to develop various levels of training at the sites under consideration.

Trainers may fail to achieve goals for many reasons, one of those reasons often being having little or no understanding of the training plan, or how that training plan has been devised. Hence, given the specificity of the subject matter, it may be advisable to review some key definitions before further exploring HZS' approach.

The first question we should ask ourselves is "what is training?". There are many documented answers, and each trainer will have his/her owns. However, the definition below is widely accepted, according to which training is [3]:

> An activity that aims at imparting the specific knowledge, skills and/or inculcate appropriate attitudes required by an individual in order to adequately perform a task or job[2]

The purpose of any training plan is to deliver results. Learning objectives state those results, and describe what people will be able to do as a result of training. They are learner-centred, and not trainer-centred. As a matter of fact, at HZS, the trainer's primary role is that of a guide in the learning process, and not just an expert imparting information to passive learners.

Another key aspect to be mentioned is the Training Needs Analysis or TNA, with its definition being [3]:

> A structured survey and analysis of training requirements arising as a result of new equipment procurement, doctrinal change, or changes to legislation, including a comparison of different training methods and equipment, with a view to recommending the optimum training system for maximum cost-effectiveness. It is a highly flexible procedure with the choice of supporting tools and techniques varying between projects. In all cases, however, a TNA is a product-based, iterative process, providing an audit trail for all decisions.

For HZS, the importance of needs analysis for training at nuclear disaster sites – and especially Chernobyl – was born out by client requirements, and by the need to recognise what may well happen again. Tragically, such importance was also confirmed by the Fukishima disaster, in April 2011. Another trigger for TNA is the evolution of 'threats', with the use of dirty bombs and radioactive dispersal devices having been much documented. Training at nuclear disaster sites also entails significant advantages for radiation mapping, environmental effect assessment, and spread and contamination control, just to mention a few areas.

Fundamental to the philosophy of need is that these steps constitute a cycle, with evaluation bringing about continuous reassessment of the needs, and the consequent refinement of the training given. This is key to Nuclear Emergency Training (NET) [6].

The NET course – like any course within the HZS training repertoire – follows a training strategy whereby it can be delivered either individually or collectively, with HZS training strategy including a pre-requisite, namely that trainees must give proof of having attended minimum 30-h radiation protection basic training [4]. This is important for two reasons:

1. The exclusion zone is not a place for "radiation rookies."
2. In order to make the most of the time spent in the zone, not wasting it on matters that can be dealt with in a classroom.

The course is designed for trainees to become aware of emergency measures after a nuclear accident. The training content includes:

1. Radiation Safety
2. Consequences of a nuclear accident
3. Consequence management after a nuclear accident

---

[2]JSP 822 – Part 2 – Training & Education Glossary. United Kingdom M.O.D.

4. Chernobyl 1986- Causes and effects
5. Comparison of Chernobyl/Fukishima
6. Operating in a radiological contaminated area
7. Protection against radioactive open sources on terrain and infrastructure.
8. Detection of radioactive sources in a contaminated environment.
9. Sampling
10. Decontamination

These objectives are not indeed set in stone. As a matter of fact, if there is a particular area which clients may want to explore, it can be investigated within the boundaries of safety and dose management. The objectives are devised based on the need to confirm previous knowledge and provide a learning experience using Bloom's Taxonomy [5].

The objectives have been divided into three "domains", namely cognitive, affective, and psychomotor (sometimes loosely described as "knowing/head", "feeling/heart" and "doing/hands" respectively) ones, thus allowing designing a more holistic and flexible training program.

As access to Chernobyl and to the exclusion zone is restricted, how is it indeed possible to reach the aforesaid objectives? First of all, anyone – as long as they are over 18 years of age and in good health – can go to Chernobyl; as a matter of fact, the BBC's motor show Top Gear did, and tourists can also book tours. The difference is that those people carry cameras, while HZS trainers and trainees carry detection devices and wear protection equipment and gear. However, despite the possibility to be authorised to access the exclusion zone, the Ukrainian authorities prefer to keep close control on whoever explores the areas carrying detection equipment. This is why HZS has entered into an agreement the Chornobyl Centre [7]. This unique partnership has allowed HZS to conduct its own radiation mapping of the ghost town of Pripyat, in order to ensure that the dose limits in HZS' training area fall within safe ranges. The partnership also allows access to the Chernobyl Power plant and the Exclusion Zone by the workers train, which leaves from the city of Slavutytch, crosses the Border with Belarus, and goes directly to the plant. Slavutytch is mostly home to the disaster survivors and their descendants, who had to be relocated as a result of the accident. The journey is quite interesting, as our trainees are encouraged to watch the rise and fall of background readings as the train travels through a small area that is believed to be in the downwind hazard. The Chernobyl railway station also provides some innovative training options in the areas of contamination control, decontamination, whole-body monitoring, and even indoor training at the station.

Generally speaking, first responders and military do not like maths; so dose monitoring is key. Trainees are made aware of what their instruments are actually telling them. Pripyat exhibits not only radioactive hazards, but also many other hazards associated with any concrete monolith that may have been left in the contaminated area for 20 years. Failing structures, open manholes, and loose footings are now remnants of a once vibrant city that had to be evacuated in a hurry. Of course, such hazards differ significantly from the hazards one would expect after around a 1 to 3-day time scale, but whether it is Pripyat now or Pripyat the day after the accident, radiation monitoring itself intensifies those hazards. Why? Give a man a tool, and he will complete a job. However, unless he already understands that tool and uses it effectively, he will lose focus on that job. So, HZS training makes trainees aware of what their instruments are telling them, but it also highlights the 'detector compass'. Eyes down, mind focused on the tool and the numbers it gives, oops, I have just fallen down a hole!!! These are key lessons that can only be learnt through field training in a contaminated environment, when radiation plays the multi-directional battle, and this is essential to the "train as you fight".

4.4 Radiation Level Measurement in Pripyat

Exact Dose Rate measurement in Pripyat was necessary to establish a safe working environment, based on Dose/ Time calculations.

4.4.1 The Initial Phase

The abandoned town of Pripyat was decontaminated in the months after the Chernobyl Nuclear Power Plant accident. Due to the time elapsed and to the natural effect of 25-years passing seasons, there are rather low levels of background radiation. Radioactive particles travel about 1 cm per year into the earth; therefore, in the bushes and on the meadows the dose rates are, in most cases, reasonably low, except in those places where highly radioactive waste was buried. But they are not included in HZS training areas.

Nevertheless, there are still some places that show high dose rates, which are still too high to ensure safe working conditions. Those are either original hotspots that had not been decontaminated, or places where hot particles have been accumulating over the years, or areas where moss is transporting radioactive particles out of the ground to the surface.

To ensure safe working conditions based on Dosimetry standards, the maximum dose was fixed at 10 μSv per day, for 3 working days.

Another challenge to safe working conditions arose during the measurements carried out around open manholes. During the decontamination process, they were all opened and they were never closed afterwards (Fig. 4.1).

4.4.2 Generating the Radiation Map

In order to ensure reliable and quantitatively significant radiation data, a FLIR Identifinder was used, in combination with a LaBr detector. The Identifinder is able to log the dose rate per second. For the actual position, a GPS was used with

Fig. 4.1 Manholes in Pripyat ©HZS

external U-Block SirfIII antenna, which was set to log the position per second. On both devices, the real time clocks where synchronized. For the measurements, the Identifinder was carried 1 m above the ground and the GPS antenna was positioned above shoulder level.

The training area was divided into three parts, which were covered in 2 days. The main constraints were due to the fact that it was necessary to stick to the timetable of the power plant workers train – thus having a specific amount of time available every day – and the fact that the area had to be covered on foot in order to collect accurate data.

For the measurements, the logging of the GPS and the Identifinder were started and stopped at the same time. After returning to the Chornobyl Centre, the log-files of the two devices were downloaded and converted, so to become readable by a table calculation software. Then, the two files were merged into one, containing the position, the date, time, and the dose rate (Table 4.2).

Then, this file was converted by a GIS program to produce colour tracks, showing the dose rate from background reading in blue, and maximum dose rate in read. The map used was taken from Open Street Map and corrected with the actual GPS information (Table 4.3).

As a 10 μSv per working day maximum dose had been fixed, a calculation base was then obtained showing in which part of our training area it was possible to work, and for how long, without exceeding the maximum daily dose. According to

Table 4.2 HZS Logfile

Datei Bearbeiten Format Ansicht ?

51.40051446°N	30.06349330°E	18. Apr 12	09:37:35.000	0,56
51.40051413°N	30.06349314°E	18. Apr 12	09:37:37.000	0,564
51.40051128°N	30.06349733°E	18. Apr 12	09:37:39.000	0,542
51.40052536°N	30.06349598°E	18. Apr 12	09:37:41.000	0,547
51.40052787°N	30.06348894°E	18. Apr 12	09:37:43.000	0,538
51.40052838°N	30.06347671°E	18. Apr 12	09:37:45.000	0,542
51.40054497°N	30.06347151°E	18. Apr 12	09:37:47.000	0,547
51.40054866°N	30.06347469°E	18. Apr 12	09:37:49.000	0,547
51.40055168°N	30.06347201°E	18. Apr 12	09:37:51.000	0,551
51.40054062°N	30.06346916°E	18. Apr 12	09:37:53.000	0,562
51.40053609°N	30.06346682°E	18. Apr 12	09:37:55.000	0,567
51.40053274°N	30.06346615°E	18. Apr 12	09:37:57.000	0,552
51.40052083°N	30.06346799°E	18. Apr 12	09:37:59.000	0,568
51.40052704°N	30.06347050°E	18. Apr 12	09:38:01.000	0,563
51.40054665°N	30.06346648°E	18. Apr 12	09:38:03.000	0,564
51.40055269°N	30.06346011°E	18. Apr 12	09:38:05.000	0,554
51.40057012°N	30.06345978°E	18. Apr 12	09:38:07.000	0,583
51.40057213°N	30.06345055°E	18. Apr 12	09:38:09.000	0,59
51.40061823°N	30.06351560°E	18. Apr 12	09:38:11.000	0,601

Table 4.3 Screenshot of the position record of other hazards

C73 fx

	A	B	C	D	E
1	Basic coordinates		51.24.xx N	30.03.xx E	
2	Number	Type of danger	Coord.	Road side	
3	1	Hole	3		
4			49		
5	2	Deep hole	4	Left side	
6		Other 2 with bad covers	47		
7	3	Hole	4	Right side	
8		Another suspect cover	43		
9	4	Hole	5	Right	

the above-mentioned thresholds and calculations, the maximum time to be spent in the exclusion zone is 4 h (set by the timetable of the train between the Chernobyl Centre and the Power Plant). As around 1 h is necessary for preparation and post-training decontamination, the adopted calculation base was 3 h. Hence, the average dose rate should not exceed 3 μSv/h. The dose rate on the decontaminated roads is just a little higher than average background levels in non-contaminated areas. The hot spots show significantly higher levels, but those areas are quite small. A colour code was developed on the radiation map, showing for how long it is possible to work and train in each area:

- Blue: No time limit
- Green: 3 h maximum time
- Yellow to Red: immediate return after 1 h (safe direct path return to stay below 10 μSv) (Fig. 4.2).

Fig. 4.2 Dose rates – HZS colour coded map showing the working time in blue to red ©HZS

When using this map, it is also important to consider the time one needs to return to the vehicle. Vehicles can travel on most roads safely.

The areas showing higher dose rates – which cannot ensure safe working conditions – are blanked out and not shown on the map. Remaining inside the coloured areas is therefore mandatory.

It was also decided to keep the colour track and not to produce a heat map showing average dose rates, because of the thick growing trees and bushes where it had not been possible to perform any measurements. In those areas, there can still be some hotspots it was not possible to measure and record any data. A heat map would give a rough idea of the areas where no measurement was taken.

4.4.3 *Other Hazards*

The buildings of Pripyat are dangerous structures, which can collapse any time. Entering those buildings is forbidden.

There are also several open manholes, whose mapping was started by taking GPS Positions and pictures. However, as there seem to be countless of such dangerous spots, this reconnaissance activity was stopped after the first day. When working and exercising in Pripyat, one should always look where he/she is putting his/her feet (see Fig. 4.1).

Fig. 4.3 Operator in the exclusion zone ©HZS

4.5 Conclusions

This training in the exclusion zone of Chernobyl is a competence based training that enhances skills and confidence of operators in their procedures and equipment in a real and live environment.

HZS has proven – by extensive field surveys – that this can be done by NEVER exceeding a daily dose of 10 μSv.

It is therefore very evident that this training enables first responders to SAFELY

- practice techniques and procedures in emergencies involving radioactive material
- apply joint and combined doctrine and tactics in a real environment.

Conclusion: training of first responders at sites of nuclear disasters enhances their preparedness for incidents involving nuclear and radioactive material (Fig. 4.3).

References

1. Borovoy AA, Begichev SN, Burlakov EV, Gavrilov SL, Dovbeyko AA, Levina LA, Markushev VM, Marchenko AE, Stroganov AA, Tataurov AL (1990) Fuel of the ChNPP reactor unit # 4 (brief reference book). Preprint. Kurchatov Atomic Energy Institute, Moscow, IAE-5268/3

2. Pazukhin (1999) Lava-like fuel containing masses of the ChNPP reactor unit # 4. Physical and chemical properties, scenario of formation, environmental impact. Post-Doctorate thesis, Specialty # 21.06.01. Chernobyl, manuscript, 1999
3. Ministry of Defense (2012) JSP 822 – Part 2 – Training & Education Glossary. United Kingdom M.O.D
4. Ministry for Agriculture and Environment (2006) Austrian norm on radiation training of intervention personnel in case of radiological emergency situations ÖNORM S 5207
5. Bloom BS, Engelhart MD, Furst EJ, Hill WH, Krathwohl DR (eds) (1956) Taxonomy of educational objectives – the classification of educational goals – handbook 1: cognitive domain. Longmans, Green & Co. Ltd, London
6. Seibersdorf Laboratories – Seibersdorf Academy – International Training Programme. https://www.seibersdorf-laboratories.at/en/products/international-training-programme/cbrn-training/nuclear-emergency-training-net.html
7. Chornobyl Center for Nuclear Safety, Radioactive Waste and Radioecology. http://www.chornobyl.net/en/

Chapter 5
Urban Response to CBRNe. Multi-service, Multi-jurisdictional, Response to Terror Events

Gilbert J. Gelineau

Abstract Terrorist attacks occur in the urban environment to maximize fear and confusion. The "urban first responder" is the critical asset in minimizing the effects of an attack. Urban first responders are first on scene, and first to extract and treat casualties. They are the first wave in protection of the public and infrastructure.

Special Operations Task Forces (SOTF) are the second wave of response. These task forces usually comprise urban police, fire and paramedic personnel with specialized training in Chemical, Biological, Radiological, Nuclear and explosive (CBRNe) identification and containment, including treatment of the contaminated and injured.

Federal policing and military units are the third wave of response. They respond at the request of the state and/or province, and are highly trained and effective specialists with advanced equipment and capabilities.

All levels of government are responsible to effectively minimize the effects of terror events to the public, infrastructure and country. A high level of capability has been developed, unfortunately within silos, to respond to a CBRNe attack but there are deficiencies within the three waves that will prevent us from staging an optimum response.

First responders are a critical key to an effective response. Indeed, the initial actions of the first responder will often be the defining factor in the outcome. Linkages and communications between the three waves of the response are also critical. Federal and military teams are certainly necessary for a full response but they must develop relationships with the first and second waves in order to be successful. Therefore, the complete response approach, pulling together the three waves of response, is critical to our public safety mission, which is to effectively coordinate and apply the appropriate resources at the appropriate time of an event.

G.J. Gelineau (✉)
Hazard Assessments & Emergency Management Solutions, Ottawa, ON, Canada
e-mail: gilbert@haemsolutions.com

S. Apikyan, D. Diamond (eds.), *Nuclear Terrorism and National Preparedness*,
NATO Science for Peace and Security Series B: Physics and Biophysics,
DOI 10.1007/978-94-017-9891-4_5

5.1 Introduction

Chemical, Biological, Radiological, Nuclear and explosive (CBRNe) events are an ever increasing concern in todays geopolitical environment. The focus on protection of the top governmental and institutional terror targets has increased their resilience and reduced their threat potential. However, the result is that softer targets are now at increased risk due to accessibility. Public gathering places, shopping malls, etc., anywhere there is free open access with large groups of the public, fall into this scenario. For example, consider the Westgate shopping mall attack in Nairobi in 2013. This effect puts "urban first responders" on the front line of (CBRNe) events. Specialized teams and taskforces provide subsequent waves of response and these have their own governance, operating systems and policies. This paper will explore first response actions and capabilities as well as interdependencies within the additional levels of response necessary to mitigate and minimize a terrorist event in an urban environment.

5.2 Terrorist Event

Imagine a soft target terrorist attack. It's a hot summer day; the local mall in downtown "Publicville" is full of people. People are shopping and just taking a break from the heat and humidity outside. The time is 12:40 pm and the food court is packed with shoppers and people grabbing some lunch. There are approximately 150–250 shoppers, students and elderly eating and gathering in a relatively enclosed area. At 12:45 pm, there is a large explosion in the centre of the food court. People within 10 m of the explosion are killed instantly; the blast wave blows out windows but does not damage the main structure of the building. There are 20 immediate fatalities plus 50 non-ambulatory casualties, with the balance of people fleeing the area.

5.3 Multiple Response Waves

The first response wave consists of municipal fire, police and paramedic personnel. Upon arrival, units are overwhelmed by the number of casualties and call for additional back up and elevated priority status. Police secure the perimeter while firefighters remove casualties to an initial triage location established by paramedics. Paramedics triage, treat and transport high-risk casualties to local hospitals. The next 1–1.5 h comprise the first wave of an urban response to a terrorist event.

The second wave includes municipal or regional Special Operations Task Forces (SOTF). The SOTF bring additional skills, personnel and equipment to the scene. The first and second waves overlap and "should" flow seamlessly, with common command and control systems. The second wave will identify and contain any contaminants, and begin decontamination for all victims and responders. The third wave of response, typically arriving 4–8 h after the event, comprises provincial/federal or military teams that assist, if required, with the response logistics, as well as begin to investigate and gather evidence.

5.4 Multiple Response Waves – The Problem

The public expects all levels of government to quickly and effectively respond to save lives and mitigate events where people have been injured, need help and require protection. Many agencies, from first responders to the scientific community, operate under misconceptions of response capability and organizational readiness. These are due to "silo'ed" mentalities. The gaps in optimal effectiveness are not clearly identifiable until multi-service training and realistic scenario role-playing exercises are performed, or an actual real terror event occurs, and the inquest report identifies deficiencies. Multi-service and multi-jurisdictional training and exercises rarely occur in most urban jurisdictions. When they do take place, competing cultural and fiscal agendas can and do degrade the effectiveness of the training and exercise. The three waves of a response will proceed in any event and, unless high quality preparedness via response objectives, capability verification and common command/control systems are in place, the potential of a successful or optimal outcome are diminished. In the case of CBRNe events, we, the responders, do not get a second chance. The first 30 min of a CBRNe response can and will make the difference in physical and mental human casualties, and in infrastructure loss or damage, not to mention the psychological effects on society.

5.5 Critical Policy and Linkages

What is required to ensure that, when an event occurs, all waves of response perform at maximum capability and in concert to minimize human suffering and damage to infrastructure? Firstly, CBRNe awareness training for municipal first responders is critical. In current best practice, risk-based evaluations of low-frequency high-risk category events should be used as the basis for regular training, as recent operational memory is required in order to maintain a high level of response capability. SOTF specializing in CBRNe tend to be focused primarily on terror events *per se* and do maintain operational capability; however, they lack linkages and coordinated policies with first responders.

Let's take the above terrorist attack to the next level. The explosion was a dissemination method for a radioactive isotope, that is, a so-called "dirty bomb". If the first responders lack the capability to identify the presence of radiation, hopefully the second wave of response will both have and use the capability. Unless a clear policy is established *apriori* that states that the identification and elimination of possible CBRNe elements is a key objective for this second response wave, the identification and elimination of these elements may not occur until late in the event, thus leading to many more casualties. For example, emergency personnel and ambulances may spread contamination to multiple hospitals and areas within the city.

Therefore, coordinated policies, response plans, networks and chain of command must be established (as part of the overall CBRNe strategy) in order minimize damage and public panic.

Two channels of communication are necessary to ensure that all waves of a CBRNe response are effective. The first is coordinated policies and procedures. All waves and levels of response in a given city or municipality, where a high threat level exists (either through institutional or governmental targets, national assets, recreational arenas or public transit systems), must establish a common command and control structure, based on frequent face-to-face relationships. Indeed, personal networks must be established or policies may fail. When governmental agencies and response agencies rely on policy alone, there is no back up for the failure of policy. Therefore, personal networks (or "back channels" as some agencies refer to them) and relationships are the second critical channel of communication. Policy and personal networks back each other up: when policy fails, personal networks will pick up the slack and fill the gap; where personal networks fail, policy will at least dictate appropriate actions. But both are essential for a truly coordinated effective response.

5.6 Summary

Terrorist events will continue to occur, as was recently evidenced in my hometown of Ottawa. Although the outcome was satisfactory, optimal policies and critical linkages were shown not to be in place. The three waves of response must be coordinated with carefully considered policies and procedures, and re-enacted with all response personnel in realistic terror event scenarios. Additionally, personal networks and relationships between the response teams must be encouraged. These gaps must be identified, rectified and maintained.

We get one chance during an event to do our jobs properly to effectively minimize casualties, public panic and damage to infrastructure. Most agencies believe they are prepared, trained and practiced for a terrorist event. However, I respectfully believe that, unless all waves of a response act as a coordinated single unit, we are not prepared and will not perform at our best. The public expects and deserves our full commitment on this.

Chapter 6
Armed Forces Radiobiology Research Institute: R&D, Training, Crisis Response. Accomplishments, Opportunities and Policy Questions

Mark H. Whitnall

Abstract The Armed Forces Radiobiology Research Institute (AFRRI) mission is to preserve and protect the health and performance of U.S. military personnel through research and training that advance understanding of the effects of ionizing radiation. This mission encompasses (1) basic and applied research to identify and develop measures to prevent, assess and treat radiation injury; (2) education and training to maintain a pool of qualified radiation biologists, health care providers, disaster preparedness personnel and operational planners; and (3) maintenance of an advisory team ready to be activated in the event of radiological crises and consequence management missions. Products resulting from the AFRRI program will prevent, mitigate and treat radiation injury and provide guidance for medical management. These products will expand the ability of warfighters to accomplish missions and will improve morale. AFRRI is funded to identify and develop products to the point where they are poised for advanced development. A number of countermeasure candidates have been developed with efficacies that would save tens of thousands of people in a mass casualty scenario. Examples of current research findings are: (1) Effect of radiation quality on countermeasure efficacy; (2) Accelerated hematopoietic syndrome in a minipig model; (3) Roles of REDD1 and miRNA30c in radiation injury and (4) multiparameter biodosimetry in a nonhuman primate model. Policy questions include the timing of administration relative to irradiation, the possibility of far-forward fielding, the administration route, and whether the priority is mass casualty or small-scale scenarios (which affects the availability of medical care).

The views expressed do not necessarily represent the Armed Forces Radiobiology Research Institute, the Uniformed Services University of the Health Sciences, or the Department of Defense.

M.H. Whitnall (✉)
Armed Forces Radiobiology Research Institute, Uniformed Services University of the Health Sciences, Bethesda, MD, USA
e-mail: mark.whitnall@usuhs.edu

S. Apikyan, D. Diamond (eds.), *Nuclear Terrorism and National Preparedness*, NATO Science for Peace and Security Series B: Physics and Biophysics,
DOI 10.1007/978-94-017-9891-4_6

6.1 Introduction

The AFRRI mission is to preserve and protect the health and performance of U.S. military personnel through research and training that advance understanding of the effects of ionizing radiation. This mission encompasses (1) basic and applied research to identify and develop measures to prevent, assess and treat radiation injury; (2) education and training to maintain a pool of qualified radiation biologists, health care providers, disaster preparedness personnel and operational planners; and (3) maintenance of an advisory team ready to be activated in the event of radiological crises and consequence management missions. Products resulting from the AFRRI program will prevent, mitigate and treat radiation injury and provide guidance for medical management. These products will expand the ability of warfighters to accomplish missions and will improve morale.

6.2 AFRRI History

AFRRI was founded in 1961 with the purpose of understanding and counteracting the acute effects of high-dose external radiation on biological systems and upon developing casualty criteria. With the end of the Cold War in 1991, AFRRI refocused its research based on new military requirements and recognized its responsibility to respond to radiation accidents and other incidents, providing medical and health physics expertise to military commands. By 1998 new nuclear threats were emerging in India, Pakistan, Iran, and North Korea. The terrorist attacks on September 11, 2001 led to a renewed appreciation of nuclear/radiological threats and to a redefinition of the role of AFRRI in homeland security. In 2001, at the request of the White House, AFRRI played a major role in determining the radiation kill curves necessary to inactivate anthrax spores that had been sent to members of Congress through the U.S. Postal Service.

AFRRI is funded for research and initial development of product candidates to the point where candidates are poised for advanced development (i.e., around the time of an Investigational New Drug (IND) application in the case of drugs). AFRRI is not funded for advanced development, although the Institute does assist with some studies in large animals, has instituted a Good Laboratory Practice (GLP) program, and provides data to product sponsors for Food and Drug Administration (FDA) submissions. Advanced development of AFRRI products has been carried out by small private companies funded by government agencies or venture capital. AFRRI has developed a portfolio of products ready for advanced development. Further development of these products will depend on decisions by the DOD as to the priority that should be given to radiation countermeasures and radiation diagnostic platforms.

In addition to appropriated funds directed to AFRRI, more than 80 % of AFRRI's research funding comes from competitive extramural grants, agreements and contracts. Research areas range from cellular and molecular exploration of radiation injury and drug mechanisms and biomarkers, to evaluation of platforms for biodosimetry in the field and testing of countermeasures in large animal models.

6.3 Research and Development Accomplishments

6.3.1 Biodosimetry

Radiation injury diagnostics (biodosimetry) is a vital part of the response to a radiological disaster. In a mass casualty scenario, over 100,000 people may be suspected of receiving medically significant doses of ionizing radiation [1]. It would be prohibitive in terms of resources to administer radiation countermeasures to all of them, and this would lead to significant unnecessary side effects from the treatments. Therefore, it is necessary to have a method of assessing the severity of radiation damage to guide medical treatment. The gold standard is the cytogenetic approach of quantifying chromosomal dicentrics in metaphase spreads of lymphocytes [2].

However, the dicentric assay suffers from low throughput that limits its usefulness in a mass casualty scenario. AFRRI adapted and validated alternative cytogenetic chromosome aberration assays with higher throughput [i.e., rapid interphase chromosome aberration assay (RICA), gamma H2AX assays], and established methods to assess partial-body and high-dose radiation exposures [3–5].

To explore other methods with higher throughput, AFRRI adopted real-time PCR methodologies for quantitative measurement of radiation-induced DNA mutation and gene expression biomarkers for use in radiation dose assessment. AFRRI was the first to apply the use of multiple protein targets for radiation dose assessment to predict an acute radiation syndrome (ARS) outcome [6].

AFRRI demonstrated the enhanced radiation diagnostic utility of using combined hematology and blood protein biomarkers (multiparametric biodosimetry), employing mouse and NHP radiation models [7].

6.3.2 Mechanisms of Radiation Injury

AFRRI has a vigorous program investigating cellular and molecular mechanisms of radiation injury and interactions of radiation injury with infectious agents in vivo and in vitro. This work is required for drug licensure: a detailed mechanistic understanding of radiation injury and drug efficacy is required by the FDA for approval of radiation countermeasures under the Animal Rule. The Animal Rule was established by the FDA for consideration of drugs like radiation countermeasures whose efficacy cannot be tested in humans, but must be demonstrated in animal models.

Radiation mechanism studies are also crucial for the assessment of radiation risk standards. The event at Fukushima in 2011 and the U.S. military's involvement afterward suggested that greater knowledge of radiation mechanisms is warranted as this will better assist with improving risk standard analysis. AFRRI has expanded its support of low dose radiation studies to specifically aid in providing data that will improve radiation risk assessment.

In vivo studies have made important advances in establishing new models of radiation exposure, refining previous models, and discovering new important physiological responses to radiation exposure. AFRRI designed a behavioral test battery to assess radiation-induced performance decrement in an animal model.

AFRRI developed the Gottingen minipig as an alternative large animal model to test radiation countermeasures, complementary to canines and NHP. The model has proven successful in terms of mimicking the human hematopoietic and GI subsyndromes of ARS, and demonstrating the efficacy of the standard ARS countermeasure, G-CSF (Neupogen®) [8, 9]. In the process of developing this model, AFRRI recently demonstrated that the minipig displays an accelerated hematopoietic syndrome at radiation doses between those that cause the hematopoietic and GI syndromes. This was the first demonstration of the accelerated ARS subsyndrome in a large animal model. Moreover, AFRRI demonstrated in this model that the systemic inflammatory response syndrome (SIRS) is an important component of ARS [10]. This aspect of ARS had been neglected up to now.

AFRRI documented responses of various classes of enteric bacteria to whole animal irradiation, providing guidance as to the types of antimicrobials that would be optimal for control of specific classes of microbes [11].

A large proportion of radiation casualties from detonation of a radiological/nuclear weapon will suffer combined injuries: radiation combined with wounds, burns, or other insults. AFRRI demonstrated a remarkable synergy between radiation and other insults in inducing mortality, and has elucidated important mechanisms contributing to the synergy [12, 13]. This led to identification of countermeasures effective against combined injury (see countermeasures section below).

It was shown that endogenous REDD1 levels are very low in human hematopoietic progenitor CD34+ cells regardless of radiation, but highly expressed in differentiated hematopoietic cells in response to radiation, which might be associated with radiation tolerance of the latter cells. Pre-miR-30c transfection suppressed REDD1 expression in 14 day cultured CD34+ cells and osteoblast niche cells and resulted in osteoblast cell death. In contrast, inhibition of miR-30c expression significantly enhanced clonogenicity in CD34+ cells [14].

6.3.3 Radiation Countermeasures

A major thrust of AFRRI's research has been the identification and initial development of radiation countermeasures. In this area, AFRRI has demonstrated remarkable productivity with limited budgets. The central thrust of AFRRI's program is to develop agents that will promote survival after exposure to external penetrating ionizing radiation. (Note: the FDA-approved radionuclide blocking or decorporation agents potassium iodide (KI), Prussian blue and DTPA do not address the issue of external penetrating radiation, the major cause of radiation casualties after a nuclear detonation.)

AFRRI brought a panel of promising radiation countermeasures to the point where advanced development by other agencies or private companies is appropriate. Private companies require government funding for advanced development of these agents. The DOD is assessing the priority of advanced development of radiation countermeasures relative to countermeasures to chemical and biological agents.

AFRRI performed the earliest work on cytokines and growth factors as radiation countermeasures, leading to acceptance of G-CSF (Neupogen®) as the standard radiation countermeasure.

In the 1990s, there was some skepticism about the ability to identify an effective radiation countermeasure with low toxicity. In 2000, AFRRI introduced 5-androstenediol (5-AED), a countermeasure with low toxicity that improved survival after radiation exposure [15, 16]. This became the prototype for a new generation of practical radiation countermeasures appropriate for advanced development.

The success with 5-AED was followed by another independently identified and developed AFRRI countermeasure: genistein, a soy isoflavone. Like 5-AED, genistein displayed low toxicity and enhanced survival in irradiated animals [17, 18].

The development of a radiation countermeasure that can prevent radiation injury if administered before exposure and mitigate radiation injury when delivered after radiation has been pioneered at AFRRI. Phenylbutyrate is a drug with low toxicity that can effectively prevent and mitigate ARS. It has also been shown to be effective when delivered orally and is well tolerated at high doses. Phenylbutyrate is the first dual protection countermeasure as it can also reduce the risk of radiation leukemia in an animal model [19].

AFRRI discovered that the radioprotective efficacy of 5-AED, CBLB502, tocopherol succinate, delta-tocotrienol, and gamma-tocotrienol is mediated through G-CSF induction. Administration of G-CSF antibody completely abrogated the radioprotective efficacy of these radiation countermeasures [15, 20, 21].

AFRRI identified G-CSF and interleukin-6 (IL-6) levels, along with complete blood counts (CBC), as efficacy biomarkers for CBLB502. This is important in estimating the effective dose in humans based on animal studies and blood assays during safety trials in humans [21].

Using the mouse model, it was demonstrated that myeloid progenitors can mitigate radiation injury when administered several days after radiation exposure. This timing of drug administration would be extremely valuable during a radiological/nuclear disaster scenario [22].

Recent results from AFRRI indicate tocols can be used to mobilize hematopoietic progenitors based on their ability to induce high levels of G-CSF. These agents may be able to replace the use of G-CSF in the clinic for mobilizing progenitors for stem cell transplants. Tocols are inexpensive compared to G-CSF, have fewer side effects, and can be stored at ambient temperatures, unlike G-CSF [23].

Almost all radiation countermeasure research has taken place in the context of photon radiation (gamma-rays and X-rays). However, radiation casualties from nuclear criticalities are exposed to a type of radiation with different mechanisms of injury: mixed neutron/gamma fields. Different mechanisms of injury affect countermeasure efficacy, depending on the mechanism of action of the drug. AFRRI demonstrated that some countermeasures effective against photons (e.g., thrombopoietin mimetics) are ineffective against mixed fields, while others (e.g., Neupogen®, CDX-301) are effective against both qualities of radiation [24].

For combined injury (exposure to ionizing radiation combined with wound or burn injury), AFRRI demonstrated that some countermeasures effective against radiation alone were not effective against combined injury. However, some countermeasures such as ghrelin [25] and ciprofloxacin (by mechanisms other than antimicrobial activity) [26] are more effective against combined injury than against radiation injury alone, which may provide clues as to the mechanisms of synergy between radiation and other insults.

6.4 Products Completed/Fielded

AFRRI developed and deployed a computer-based software diagnostic tool, the Biodosimetry Assessment Tool or BAT, for use by health-care providers early after a radiation incident. This tool assists providers in identifying individuals with significant radiation exposures and in making appropriate treatment decisions. AFRRI transitioned an encrypted version of BAT (eBAT) via Medical Communications for Combat Casualty Care (MC4) for fielding in the desktop version of the Electronic Medical Record (EMR) software suite.

AFRRI developed and released a prototype of AFRRI's First-responders Radiological Assessment Triage (for Windows) software diagnostic tool to enable first responders to triage suspected radiation casualties based on the initial or prodromal features listed in AFRRI's Emergency Radiation Medicine Response – Pocket Guide. This tool represents the first application that weighs responses for various diagnostic endpoints to provide an initial multiple parameter consensus "triage" dose.

AFRRI developed and deployed a Biodosimetry Worksheet, which provides places for recording facts about a case of radiation exposure, including the source and type of radiation, the extent of exposure, and the nature of the resulting injuries. Use of the Worksheet by U.S. forces is compliant with NATO STANAGs.

AFRRI's early work on cytokines and growth factors as radiation countermeasures [27, 28] led to inclusion of G-CSF (Neupogen®) in the Strategic National Stockpile. In a radiation emergency, it would be deployed under an Emergency Use Authorization. A recent FDA panel confirmed the official consensus that G-CSF is an effective radiation countermeasure and it is now licensed by the FDA for the indication of ARS.

6.5 Support Provided to Operational Units

The Medical Radiobiology Advisory Team (MRAT) provides health physics, medical, and radiobiological advice to military and civilian command and control operations worldwide in response to nuclear and radiological incidents requiring a coordinated federal response.

Through "reachback," the deployed team of radiation medicine physicians and senior health physicists can call on the knowledge and skills of radiobiologists, biodosimetrists, and other research professionals at AFRRI as well as those of other Department of Defense (DOD) response teams.

6.5.1 MRAT Services

- Augment the Defense Threat Reduction Agency (DTRA) Consequence Management Advisory Team, which provides deployable teams of 2–20 personnel who are experts in chemical, biological, radiological, nuclear, and explosives matters.
  - Military physicians advise on-scene commanders, senior officials, and local medical personnel.
  - Health physicists advise senior officials with risk assessment through analysis of plume models as well as guidance for patient, personnel, and equipment decontamination.
  - Team members interface with their counterparts from other organizations.
- Provide direct support to the National Military Command Center, the Office of the Assistant to the Secretary of Defense (Nuclear Matters), Response Task Force Commanders, and Combatant Commanders.
- Participate in the planning and execution of DOD and U.S. interagency exercises involving radiological and nuclear scenarios.
- Support two to three major command post and field training exercises each year.
- Collaborate with other operational experts to conduct graduate-level continuing education through the Medical Effects of Ionizing Radiation Course and to develop relevant information products.
  - AFRRI Handbook: Medical Management of Radiological Casualties
  - Allied Medical Publication [AMedP-6(C) Volume 1 (Nuclear)]: NATO Handbook on the Medical Aspects of NBC Defensive Operations
  - Army Field Manual (FM 4–02.283): Treatment of Nuclear and Radiological Casualties
  - DOD Manual (DOD 5130.8-M): Nuclear Weapon Accident Response Procedures (NARP, Chapters 10 and 11)
  - NATO Standardization Agreement (STANAG): Commander's Guide on the Effects from Nuclear Radiation Exposure During War

  - NATO STANAG: Guidance on the Use of Antiemetics for Radiation-Induced Nausea and Vomiting

6.5.2 Responses to Terrorism and Support to Operational Units

- Supported continuity of government operations during the anthrax attacks on the U.S. Capitol in 2001
  - Set the radiation standard for anthrax mail irradiation
  - Played a pivotal role in generation of the weaponized anthrax radiation kill curve
  - Certified Lima, OH, and Bridgeport, NJ, irradiation facilities
  - Provided quality assurance for Brentwood ClO_2 sanitation – supported continuity of government operations
- Responded to nuclear/radiation crises
  - Under the Federal Radiological Emergency Response Plan
  - Direct support to the National Military Command Center
- Provided direct support to the NORTHCOM commander
  - Military support to the Homeland Security mission
  - Involved in NORTHCOM joint exercises
- Provided direct support to the Central Command (CENTCOM) commander in Operation Enduring Freedom in multiple deployments
- During the DOD response to the Fukushima Daiichi reactor incident, Operation Tomodachi, the MRAT provided guidance and advice to U.S. Military leaders in Japan. This support helped ensure the safety of U.S. service members, family members, and civilians and supported the humanitarian relief in a coordinated effort with the Government of Japan.
- Provided guidance on treatment of personnel with embedded depleted uranium fragments or tungsten alloy fragments.

6.6 Policy Questions

- Factors affecting the likelihood of a major radiological/nuclear incident are uncertainty about North Korea and Iran, the resilience of jihadism, growing cynicism concerning nonproliferation, and the perceived utility of nuclear weapons by a number of states [29]. Given those concerns, the relative priority of funding for medical countermeasures and biodosimetry should be considered carefully.

- Any such incident would require deployment of personnel into contaminated areas, raising the question of whether radiation countermeasures to be deployed before and shortly after exposure deserve increased attention.
- Current response plans are top-down, requiring distribution of countermeasures by central authorities. Far-forward fielding (stockpiling of countermeasures at local sites) may allow earlier treatment during a radiological/nuclear disaster.
- The animal model for testing and approval of countermeasures against gastrointestinal (GI) syndrome needs to be agreed upon. This syndrome occurs after radiation doses higher than the hematopoietic syndrome, hence GI syndrome treatment must deal with both syndromes. For example the model may involve shielding of some bone marrow or administration of a countermeasure against the hematopoietic syndrome.

References

1. DiCarlo AL, Maher C, Hick JL, Hanfling D, Dainiak N, Chao N, Bader JL, Coleman CN, Weinstock DM (2011) Radiation injury after a nuclear detonation: medical consequences and the need for scarce resources allocation. Disaster Med Public Health Prep 5(Suppl 1):S32–44. doi:5/Supplement_1/S32 [pii] 10.1001/dmp.2011.17
2. Blakely WF, Carr Z, Chu MC, Dayal-Drager R, Fujimoto K, Hopmeir M, Kulka U, Lillis-Hearne P, Livingston GK, Lloyd DC, Maznyk N, Perez Mdel R, Romm H, Takashima Y, Voisin P, Wilkins RC, Yoshida MA (2009) WHO 1st consultation on the development of a global biodosimetry laboratories network for radiation emergencies (BioDoseNet). Radiat Res 171(1):127–139. doi:10.1667/RR1549.1
3. Miura T, Nakata A, Kasai K, Nakano M, Abe Y, Tsushima E, Ossetrova NI, Yoshida MA, Blakely WF (2014) A novel parameter, cell-cycle progression index, for radiation dose absorbed estimation in the premature chromosome condensation assay. Radiat Prot Dosim 159:52–60. doi:10.1093/rpd/ncu126
4. Redon CE, Nakamura AJ, Gouliaeva K, Rahman A, Blakely WF, Bonner WM (2011) Q(gamma-H2AX), an analysis method for partial-body radiation exposure using gamma-H2AX in nonhuman primate lymphocytes. Radiat Meas 46(9):877–881. doi:10.1016/j.radmeas.2011.02.017
5. Blakely WF, Sandgren DJ, Nagy V, Kim SY, Sigal GB, Ossetrova NI (2014) Further biodosimetry investigations using murine partial-body irradiation model. Radiat Prot Dosim 159(1–4):46–51. doi:10.1093/rpd/ncu127
6. Ossetrova NI, Condliffe DP, Ney PH, Krasnopolsky K, Hieber KP, Rahman A, Sandgren DJ (2014) Early-response biomarkers for assessment of radiation exposure in a mouse total-body irradiation model. Health Phys 106(6):772–786. doi:10.1097/hp.0000000000000094
7. Blakely WF, Ossetrova NI, Whitnall MH, Sandgren DJ, Krivokrysenko VI, Shakhov A, Feinstein E (2010) Multiple parameter radiation injury assessment using a nonhuman primate radiation model-biodosimetry applications. Health Phys 98(2):153–159. doi:10.1097/HP.0b013e3181b0306d
8. Elliott TB, Deutz NE, Gulani J, Koch A, Olsen CH, Christensen C, Chappell MC, Whitnall MH, Moroni M (2014) Gastrointestinal acute radiation syndrome in Gottingen minipig (*Sus scrofa domestica*): an exploratory study. Comp Med 64(6):456–463
9. Moroni M, Ngudiankama BF, Christensen C, Olsen CH, Owens R, Lombardini ED, Holt RK, Whitnall MH (2013) The Gottingen minipig is a model of the hematopoietic acute radiation syndrome: G-colony stimulating factor stimulates hematopoiesis and enhances survival from lethal total-body gamma-irradiation. Int J Radiat Oncol Biol Phys 86(5):986–992. doi:10.1016/j.ijrobp.2013.04.041

10. Moroni M, Elliott TB, Deutz NE, Olsen CH, Owens R, Christensen C, Lombardini ED, Whitnall MH (2014) Accelerated hematopoietic syndrome after radiation doses bridging hematopoietic (H-ARS) and gastrointestinal GI-ARS acute radiation syndrome: early hematological changes and systemic inflammatory response syndrome in minipig. Int J Radiat Biol 90:363–372
11. Brook I, Elliott TB, Ledney GD, Shoemaker MO, Knudson GB (2004) Management of postirradiation infection: lessons learned from animal models. Mil Med 169(3):194–197
12. Kiang JG, Ledney GD (2013) Skin injuries reduce survival and modulate corticosterone, C-reactive protein, complement component 3, IgM, and prostaglandin E 2 after whole-body reactor-produced mixed field (n + gamma-Photons) irradiation. Oxid Med Cell Longev 2013:821541. doi:10.1155/2013/821541
13. DeBell RM, Ledney GD, Snyder SL (1987) Quantification of gut injury with diamine oxidase activity: development of a fission neutron RBE and measurements with combined injury in mouse models. Radiat Res 112(3):508–516
14. Li XH, Ha CT, Fu D, Xiao M (2012) Micro-RNA30c negatively regulates REDD1 expression in human hematopoietic and osteoblast cells after gamma-irradiation. PLoS ONE 7(11):e48700. doi:10.1371/journal.pone.0048700
15. Grace MB, Singh VK, Rhee JG, Jackson WE 3rd, Kao TC, Whitnall MH (2012) 5-AED enhances survival of irradiated mice in a G-CSF-dependent manner, stimulates innate immune cell function, reduces radiation-induced DNA damage and induces genes that modulate cell cycle progression and apoptosis. J Radiat Res 53:840–853. doi:10.1093/jrr/rrs060
16. Whitnall MH, Elliott TB, Harding RA, Inal CE, Landauer MR, Wilhelmsen CL, McKinney L, Miner VL, Jackson WE 3rd, Loria RM, Ledney GD, Seed TM (2000) Androstenediol stimulates myelopoiesis and enhances resistance to infection in gamma-irradiated mice. Int J Immunopharmacol 22(1):1–14. doi:S0192056199000594 [pii]
17. Ha CT, Li XH, Fu D, Xiao M, Landauer MR (2013) Genistein nanoparticles protect mouse hematopoietic system and prevent proinflammatory factors after gamma irradiation. Radiat Res 180:316–325. doi:10.1667/RR3326.1
18. Landauer MR, Srinivasan V, Seed TM (2003) Genistein treatment protects mice from ionizing radiation injury. J Appl Toxicol 23(6):379–385
19. Miller AC, Cohen S, Stewart M, Rivas R, Lison P (2011) Radioprotection by the histone deacetylase inhibitor phenylbutyrate. Radiat Environ Biophys 50(4):585–596. doi:10.1007/s00411-011-0384-7
20. Kulkarni S, Singh PK, Ghosh SP, Posarac A, Singh VK (2013) Granulocyte colony-stimulating factor antibody abrogates radioprotective efficacy of gamma-tocotrienol, a promising radiation countermeasure. Cytokine. doi:10.1016/j.cyto.2013.03.009
21. Krivokrysenko VI, Shakhov AN, Singh VK, Bone F, Kononov Y, Shyshynova I, Cheney A, Maitra RK, Purmal A, Whitnall MH, Gudkov AV, Feinstein E (2012) Identification of granulocyte colony-stimulating factor and interleukin-6 as candidate biomarkers of CBLB502 efficacy as a medical radiation countermeasure. J Pharmacol Exp Ther 343(2):497–508. doi:10.1124/jpet.112.196071
22. Singh VK, Christensen J, Fatanmi OO, Gille D, Ducey EJ, Wise SY, Karsunky H, Sedello AK (2012) Myeloid progenitors: a radiation countermeasure that is effective when initiated days after irradiation. Radiat Res 177(6):781–791
23. Singh VK, Singh PK, Wise SY, Seed TM (2011) Mobilized progenitor cells as a bridging therapy for radiation casualties: a brief review of tocopherol succinate-based approaches. Int Immunopharmacol 11(7):842–847. doi:S1567-5769(11)00052-X [pii] 10.1016/j.intimp.2011.01.017
24. Cary LH, Ngudiankama BF, Salber RE, Ledney GD, Whitnall MH (2012) Efficacy of radiation countermeasures depends on radiation quality. Radiat Res 177:663–675
25. Kiang JG, Zhai M, Liao P-J, Elliott TB, Gorbunov NV (2014) Ghrelin therapy improves survival after whole-body ionizing irradiation combined with wound or burn: amelioration of leukocytopenia, thrombopenia, splenomegaly, and bone marrow injury. Oxid Med Cell Longev 2014:Article ID 215858

26. Kiang JG, Fukumoto R (2014) Ciprofloxacin increases survival after ionizing irradiation combined injury: gamma-h2ax formation, cytokine/chemokine, and red blood cells. Health Phys 106(6):720–726. doi:10.1097/hp.0000000000000108
27. Neta R, Vogel SN, Oppenheim JJ, Douches SD (1986) Cytokines in radioprotection. Comparison of the radioprotective effects of IL-1 to IL-6, GM-CSF and IFN-g. Lymphokine Res 5(suppl. 1):S105–S110
28. Neta R, Oppenheim JJ, Douches SD (1988) Interdependence of the radioprotective effects of human recombinant interleukin 1 alpha, tumor necrosis factor alpha, granulocyte colony-stimulating factor, and murine recombinant granulocyte-macrophage colony-stimulating factor. J Immunol 140(1):108–111
29. Allison G (2010) Nuclear disorder: surveying atomic threats. Foreign Aff 89:74–85

Part II
Detection and Protection

Chapter 7
Rad/Nuc Detection Standards and Conformity Assessment Program

Peter J. Chiaro Jr.

Abstract The Radiological and Nuclear Detection Standards and Conformity Assessment Program facilitates the establishment of radiological and environmental performance requirements through the development and publication of standards, and provides a means for controlled and repeatable testing needed to analyze a detection system's ability to function as defined by the applicable standard. Combined, these efforts help identify functional and performance limitations, provide a reliable method to ensure that limitations are eliminated or mitigated, and help ensure radiation detection systems meet the needs of the radiological and nuclear detection community. This paper will cover the basis for The Radiological and Nuclear Detection Standards and Conformity Assessment Program and the three components of the program: Rad/Nuc Detection Standards, Graduated Rad/Nuc Detector Evaluation and Reporting (GRaDER®), and Interagency Characterization and Assessment Program (ICAP).

7.1 Overview

The Radiological and Nuclear Detection Standards and Conformity Assessment Program facilitates the establishment of radiological and environmental performance requirements through the development and publication of standards, and provides a means for controlled and repeatable testing needed to analyze a detection system's ability to function as defined by the applicable standard. Combined, these efforts help identify functional and performance limitations, provide a reliable method to ensure that functional limitations are eliminated or mitigated, and help ensure radiation detection systems meet the needs of the radiological and nuclear detection community.

P.J. Chiaro Jr. (✉)
Domestic Nuclear Detection Office, US Department of Homeland Security, Washington, DC, USA
e-mail: Peter.Chiaro@hq.dhs.gov

S. Apikyan, D. Diamond (eds.), *Nuclear Terrorism and National Preparedness*, NATO Science for Peace and Security Series B: Physics and Biophysics,
DOI 10.1007/978-94-017-9891-4_7

7.2 Basis

The Radiological and Nuclear Detection Standards and Conformity Assessment Program is consistent with the Security and Accountability For Every (SAFE) Port Act of 2006, the Office of Management and Budget's (OMB) A-119 – Federal Participation in the Development and Use of Voluntary Consensus Standards and in Conformity Assessment Activities, and the National Strategy on Chemical, Biological, Radiological, Nuclear, and Explosives (CBRNE) Standards.

7.3 Rad/Nuc Detection Standards

The Rad/Nuc Detection Standards portion of the program has six components. These include supporting the development and maintenance of National and International consensus standards; developing and maintaining Government-unique, target-based Technical Capability Standards in accordance with the SAFE Port Act of 2006; and providing for the validation of rad/nuc detection standards. The typical development process for a standard is shown in Fig. 7.1. The remaining components ofthe Rad/Nuc Detection Standards portion of the program are modeling and

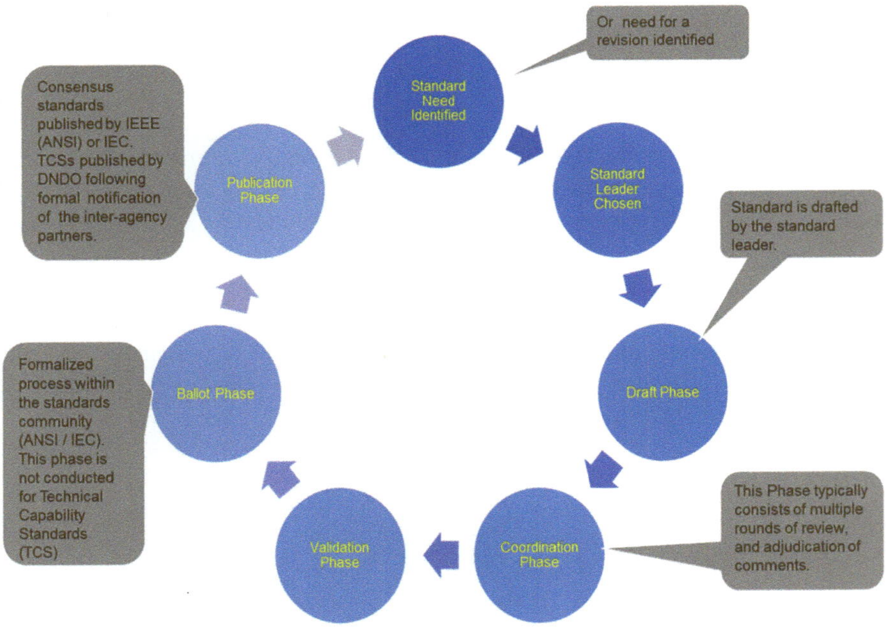

Fig. 7.1 Standards Process. This figure shows the steps associated with standards development for both consensus and government unique standards

analysis to develop requirements and verify test methods established by standards, and providing open access to standards for the preventative rad/nuc detection community.

7.4 Current Rad/Nuc Detection Standards Activities

7.4.1 National Consensus Standards

Current programmatic activities in national consensus standards include facilitating and participating in the review, revision, and publication of 6 U.S. National Standards (American National Standards Institute – ANSI):

- ANSI 42.32-Personal Radiation Detectors
- ANSI 42.33-Handheld Detectors
- ANSI 42.34-Radionuclide Identifiers
- ANSI 42.35-Radiation Portal Monitors
- ANSI 42.37-Training,
- ANSI 42.43-Mobile Detection Systems

The program is also currently developing an ASTM Guide for Standards-Based Testing. This guide will provide the Rad/Nuc detection testing community with valuable information regarding the best practices associated with standards-based testing making existing standards both more efficient and effective.

Additionally, the program is reviewing international standards, test results, and procurement requirements for applicable updates to U.S. standards when appropriate.

7.4.2 International Consensus Standards

The Program supports U.S. involvement with the International Electrotechnical Commission (IEC). The program PM is the chair of IEC Sub Committee 45B (SC45B) and a Convener and Project Leader in IEC Technical Committee 104 (TC104). The Program also supports the participation of US personnel in efforts associated with the development of performance requirements by the IAEA.

IEC SC45B prepares standards that address radiation instrumentation used for: the measurement of ionizing radiation in the workplace, the public, and in the environment for radiation protection purposes; illicit trafficking detection and identification of radionuclides; radiation-based security screening. SC45B's current activities include revising the international standards for Radionuclide Identifiers and Personal Radiation Detectors, development of a Technical Guide for the selection of radiation sources needed to test and evaluate rad/nuc detection equipment, and development of a standard for the use of spectral injection techniques for the analysis of radionuclide identification processes without the use of radiation sources.

IEC Technical Committee 104 establishes environmental condition classifications which represent the conditions to which products are most likely to be subjected while being transported, stored, installed and used. TC104 also establishes test methods intended for the preparation of product performance specifications, and transforms environmental condition classes to environmental tests. Current activities include revising standards that establish conditions associated with installed weather protected and non-weather protected electro-technical devices.

The Program also facilitates the review of draft international standards by DNDO and interagency partners.

7.4.3 Technical Capability Standards (TCS)

Technical capability standards are developed by a DNDO-led interagency group with members from other components of DHS, Department of Energy (DOE), U.S. National Laboratories, and other U.S. Federal Agencies, for example, the Nuclear Regulatory Commission and Federal Bureau of Investigation. The following Technical Capability Standards (TCSs) have been published:

- Handheld radionuclide identifiers
- Backpack-based systems
- Mobile systems

There are currently two TCSs under development. The first is an Aerial System Technical Capability Standard which establishes the radiological performance requirements for helicopter-based radiation detection systems. Validation testing of the Aerial System TCS is planned to take place in November 2014 with publication expected during 2015. The second TCS currently under development is the Advanced Radiography TCS which establishes the identification requirements for non-intrusive systems with the ability to perform materials discrimination. The coordination process for this TCS is entering its final stages and there will be a limited validation performed on this standard.

Future plans include a Maritime Rad/Nuc Detection System TCS for which the technical basis is currently being developed. Initiation of this standard is slated to begin in 2015. All Technical Capability Standards can be viewed and downloaded in PDF form at: http://www.dhs.gov/publication/dndo-technical-capability-standards.

7.4.4 Standards Validation

Validation is a process used to verify that a standard can be followed by the testing community. Validation helps to identify weaknesses in test method processes which can increase the potential for misinterpretation, produce inconsistent results, and add unnecessary costs. Standard validations involve two independent testing facilities performing each requirement and associated test method either using an applicable rad/nuc detection device or through table top exercises. Validation comments by the testing facilities are provided to the Project Leader for adjudication when needed. The program is currently validating four consensus standards, and two TCSs.

7.4.5 *Modeling and Analysis for Standards*

Modeling and analysis helps develop requirements and test methods used in standards. These activities focus on ensuring that requirements and test methods are effective at simulating operational performance requirements, and ensure that testing methods are properly set-up, performed, and recorded. This component of the Program has analyzed radiation background levels associated with the operation of mobile detection systems, evaluated techniques used to establish gamma and neutron test fields, and performed measurements of electromagnetic fields to ensure that requirements for standards are appropriate for expected use environments. Modeling and analysis is also developing an alternative electromagnetic interference test method for installed rad/nuc detection equipment and determining whether there is a cost-effective alternative neutron source that could replace the short lived ^{252}Cf.

7.4.6 *Open Access to Standards*

The Rad/Nuc Detection Standards and Conformity Assessment Program provides contracted funds to the Institute of Electrical and Electronics Engineers (IEEE) to allow open access to DNDO applicable ANSI standards. There are currently eleven standards on the open access contract:

- N42.32-2006 – American National Standard Performance Criteria for Alarming Personal Radiation Detectors for Homeland Security
- N42.33-2006 – American National Standard for Portable Radiation Detection Instrumentation for Homeland Security
- N42.34-2006 – American National Standard Performance Criteria for Hand-held Instruments for the Detection and Identification of Radionuclides
- N42.35-2006 – American National Standard Evaluation and Performance of Radiation Detection Portal Monitors for Use in Homeland Security
- N42.37-2006 – American National Standard for Training Requirements for Homeland Security Purposes Using Radiation Detection Instrumentation for Interdiction and Prevention
- N42.38-2006 – American National Standard Performance Criteria for Spectroscopy-Based Portal Monitors Used for Homeland Security

- N42.41-2007 – American National Standard Minimum Performance Criteria for Active Interrogation Systems Used for Homeland Security
- N42.42-2006 – American National Standard Data Format Standard for Radiation Detectors Used for Homeland Security
- N42.43-2006 – American National Standard Performance Criteria for Mobile and Transportable Radiation Monitors Used for Homeland Security
- N42.48-2008 – American National Standard Performance Requirements for Spectroscopic Personal Radiation Detectors (SPRDs) for Homeland Security
- N42.53-2013 – American National Standard Performance Criteria for Backpack-Based Radiation-Detection Systems Used for Homeland Security

This component of the Program helps to ensure greater acceptance and implementation of standards by the community. For 2014, at the time of this paper, 817 copies have been downloaded from the site.

7.5 Graduated Rad/Nuc Detector Evaluation and Reporting (GRaDER®)

GRaDER® provides a means to evaluate radiological and nuclear detection equipment against applicable national consensus standards and TCSs. The program identifies radiation detection products that comply with standards and satisfy Homeland Security mission requirements helping U.S. Federal, state, local, tribal and territorial agencies make more informed radiological/nuclear detector procurement decisions. Additionally, GRaDER® provides an infrastructure for the collection of high integrity test data by supporting and making use of accredited testing facilities. Go to http://www.dhs.gov/guidance-grader-program for additional information.

Currently, GRaDER® is instituting a series of reforms. We are revising the website to improve clarity and reduce the complexity of individual instrument compliance levels. GRaDER® is also working with the Homeland Security Information Network Community of Interest to improve the test result dissemination process to efficiently get test results out to the user community.

7.6 Radiological/Nuclear (R/N) Interagency Characterization and Assessment Program (ICAP)

The Domestic Nuclear Detection Office with the Department of Defense Deputy Under Secretary of the Army – Testing and Evaluation initiated R/N ICAP to develop a systematic approach for sharing test performance data that will better enable federal, state, local, and tribal government agencies to select suitable R/N detection systems. R/N ICAP provides government cost efficiencies through sharing of test infrastructure, technologies, methodologies, test events, data collection, data analysis and reporting to reduce redundant testing across Federal agencies.

R/N ICAP is currently testing Backpack Radiation Detector (BRD) systems against the BRD TCS and the neutron portion of the ANSI Backpack Standard, N42.53. Testing is taking place at Savannah River National Laboratory and includes six models from six different manufacturers. Most models have radionuclide identification capabilities and all include ^{3}He alternative neutron detectors.

R/N ICAP plans to initiate testing of Handheld Radionuclide Identifiers slated to begin in early December at Oak Ridge National Laboratory. The request for information closed on 31 October with selection to take place as soon as possible.

7.7 Conclusion

The Rad/Nuc Detection Standards and Conformity Assessment Program is very active. Each effort has ongoing and planned work. The components of the Rad/Nuc Standards effort enable DNDO to be actively engaged in standard development as technology continues to evolve benefitting the entire preventative rad/nuc community. GRaDER® is active and implementing a series of reforms to improve communication, efficiency, and effectiveness. R/N ICAP shows the value of the Interagency for conformity assessment of rad/nuc detection systems.

Chapter 8
New Opportunities of Portal Monitors with Plastic Scintillation Detectors (Asia-New. New Advanced Source Identification Algorithm)

Andrei Stavrov and Eugene Yamamoto

Abstract Radiation Portal Monitors (RPM) with plastic detectors represent the main instruments used for primary border (customs) radiation control. RPM are widely used because they are simple, reliable, relatively inexpensive and have a high sensitivity. However, experience using the RPM in various countries has revealed the systems have some grave shortcomings. There is a dramatic decrease of the probability of detection of radioactive sources under high suppression of the natural gamma background (radiation control of heavy cargoes, containers and, especially, trains). NORM (Naturally Occurring Radioactive Material) existing in objects under control trigger the so-called "nuisance alarms," requiring a secondary inspection for source verification. At a number of sites, the rate of such alarms is so high it significantly complicates the work of customs and border officers.

This paper presents a brief description of new variant of algorithm ASIA-New (New Advanced Source Identification Algorithm), which was developed by the author and based on some experimental test results. It also demonstrates the capability of a new system to overcome the shortcomings stated above. New electronics and ASIA-New enables RPM to detect radioactive sources under a high background suppression (tested at 15–30 %) and to verify the detected NORM (KCl) and the artificial isotopes (Co-57, Ba-133). New variant of ASIA is based on physical principles and does not require a lot of special tests to attain statistical data for its parameters. That is why this system can be easily installed into any RPM with plastic detectors.

This algorithm was tested for 1,395 passages of different transports (cars, trucks and trailers) without radioactive sources. Only one false alarm has been detected. It also was tested for 4,015 passages of these transports with radioactive sources of different activity (Co-57, Ba-133, Cs-137, Co-60, Ra-226, Th-232) and NORM (K-40).

A. Stavrov (✉) • E. Yamamoto
Rapiscan Systems, Inc., Torrance, CA, USA
e-mail: andreistavrov@gmail.com

S. Apikyan, D. Diamond (eds.), *Nuclear Terrorism and National Preparedness*,
NATO Science for Peace and Security Series B: Physics and Biophysics,
DOI 10.1007/978-94-017-9891-4_8

8.1 Introduction

The main instruments for the primary border (customs) radiation control are the radiation portal monitors (RPM) with plastic detectors. RPMs are widely used because these have high sensitivity, are simple, reliable and relatively inexpensive. However, the experience of the RPM usage in various countries revealed some their limitations. There is a dramatic decrease of the probability of detection of radioactive sources under high suppression of the natural gamma background (radiation control of heavy cargos, containers and, especially, trains). NORM (Naturally Occurring Radioactive Material) existing in objects under control triggers nuisance alarms, requiring secondary inspection for source verification. In a number of sites the rate of such alarms is so high that it significantly complicates the work of customs and border officers.

At the same time it should be noted that such weaknesses can be partially mitigated by various methods, e.g. auxiliary equipment application (for example, handheld instruments). The add-on equipment can be used for NORM verification in case of exceeding the alarm threshold setting. The problem of drive through detection of radiation sources at high background suppression can be addressed by use of spectrometric monitors (SRPM). However, being rather expensive, such monitors would not replace the currently installed RPM. Even for specific cases, with dozens of RPMs being installed at limited locations (i.e., sea ports), SRPMs can be used for additional inspection only in the case of alarm triggered at one or more monitors. The problem of radiation source detection with background suppression is intractable using traditional RPMs but can be solved with SRPM with mass application. Due to cost, the SRPM solution is infeasible even for developed countries. At the same time, the problem is crucially important, and even more important lately due to disturbances at the Middle East. The significance of radiation source detection can be clearly demonstrated by the following examples.

If we take ideal vehicle passage with radiation source (i.e., Co-57 of 7 sigma activity) at no background suppression (Fig. 8.1) and at background suppression (max. 3 sigma only) (Fig. 8.2), it is clear that such 7-sigma radiation source can not be detected even at the background suppression of such small value.

Further, let's analyze the effect of various factors on minimum detectable activity (MDA) of RPMs. Thus, the following variants were subject to analysis: MDA(1) – no background suppression, MDA(2) – at source suppression valued 20 sigma (40' shipping container passage, when the source is placed at the vehicle's geometrical center), and MDA(3) – at vehicle passage with the source placement at so-called "worst case" location. The comparison values are given in Table 8.1.

Reference to the Table 8.1, actual MDA value of RPM with plastic detectors, when used in field conditions, can substantially differ from the manufacturer-declared or customer-tested values. Moreover, whereas modern RPMs with plastic detectors apply the traditional algorithm based on comparison of the pre-set alarm threshold (generally, 4–6 sigma) and radiation value as may be emitted by passing vehicles, such devices are not designed to verify NORM values. At some locations

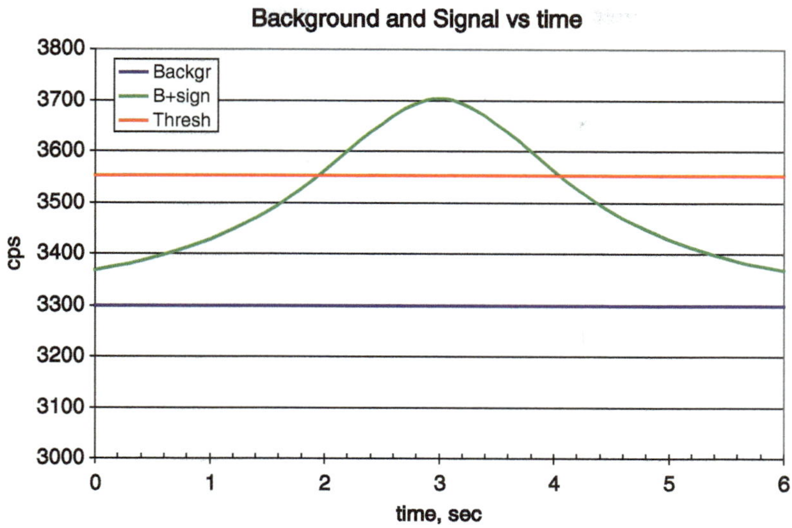

Fig. 8.1 Simulated dependence of moving source signal on time without background suppression (net signal = 7 sigma, suppression = 0 sigma, threshold TR = B + 4.4 sigma)

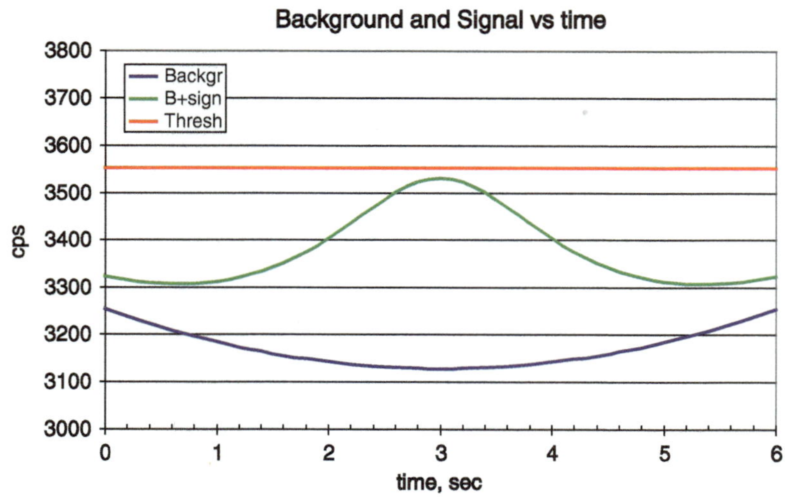

Fig. 8.2 Simulated dependence of moving source signal on time with background suppression (net signal = 7 sigma, suppression = 3 sigma, threshold TR = B + 4.4 sigma)

(as unofficially reported by monitor manufacturers), this causes numerous so-called “innocent alarms”. As a result, the monitor operator often sets a higher alarm threshold (up to 30–50 sigma) to reduce the false alarm rate. In such cases, the MDA value can exceed the declared values multifold, thus making impracticable the application of such monitor for gamma source detection.

Table 8.1 MDA change, depending on source suppression and source location in vehicle

Source	MDA(1) rel. units	MDA(2) rel. units	MDA(3) rel. units	MDA(4) rel. units
Co-57	1	7–9	10–12	16–20
Ba-133	1	5–7	8–10	14–16

8.2 ASIA Algorithm

To address these problems a new algorithm ASIA, "Advanced Source Identification Algorithm", was developed. It is necessary to clarify that the term "source identification" is principally different from the widely used term "isotope identification".

Isotope identification means the use of the spectral analysis for the determination of a specific isotope, and detectors based on inorganic scintillators are used for this purpose. Source identification means determination of a source type – artificial or natural, and plastic detectors can be used for this purpose. The ASIA algorithm uses new electronics (hardware) and a new algorithm (software) to process information from each detector. The new algorithm also uses information supplied by the traditional algorithm, with both algorithms working simultaneously and independently. With all this information available, dozens of parameters (coefficients) are calculated and analyzed, and the decision tree with many branches is generated. The final result is the message (alarm) about the detection of an artificial source. In the case of NORM detection the main alarm is not triggered, but the appropriate information is sent to the screen of a responsible operator (person) [1–4].

That previous variant of algorithm was developed using statistical analysis of numerous measurements with RPMs (in this case, by use of two types of monitors – VM-250 and TM-850, manufactured by Rapiscan Systems) during passage of various types of vehicles without radioactive sources on board and with sources – Co-57, Ba-133, Cs-137, Co-60 and with KCl as NORM. The results confirmed the efficacy of the algorithm. At the level of background suppression not exceeding 25 sigma, monitors detected various sources with activity to meet the requirements of different standards (ANSI, GOST-R), and the applicable IAEA recommendations. The algorithm also verified the NORM (K-40, 5 t of KCl) and Co-60 values (more than 0.5 mCi). However, it should be noted that this algorithm ensured excellent results for a specific monitor installed in a particular location. In case the monitor design was subjected to changes, for example using a significantly different detector size and geometry, new thresholds would need to be determined.

8.3 ASIA-New Algorithm

In order to address these issues, a new version of the algorithm has been developed – ASIA-New. Unlike the previous version, this algorithm uses a phenomenological approach and analysis of some important parameter changes during the vehicle passage through the monitor control area. In this case, the main feature of this algorithm is to make the decision when the occupancy detector is turned off, rather than on exceeding the threshold of one of parameters. In other words, prior to triggering the alarm of the presence or absence of an artificial radiation source, the entire scope of values at least 1–3 s prior to entering and exit from the monitoring control area is subject to analysis.

8.3.1 New Algorithm's Major Features

1. All algorithms are well-known to be based on use of the so-called instantaneous count rate, which means the total count rate at 200 ms intervals. At the same time, due to significant fluctuations of this parameter, smoothing is applied, where the rolling average of count rates is calculated over intervals of 1, 2, 3 or more seconds, depending on the monitor operating conditions. The smoothing value is set as a constant at virtually all types of monitors and not subject to change during operation. At the same time, the optimal value of smoothing depends on the type of vehicle under the radiation control. More specifically, on the time interval between switching on and off the occupancy detector. The shorter the interval, the smaller smoothing value is practicable to use. For example, for a passenger car, the optimum smoothing value is equal to 1 s and for long vehicle – 3 s plus.

Figure 8.3a–c shows the count rate profiles of one of the passages of a trailer with 40' shipping container, as set for instantaneous count rate, static state and with three options of smoothing applied. As can be seen from the figures, the profile obtained in the steady state is significantly different from the others not only from the various smoothing options, but also from the instantaneous count rate profile. In other words, the dynamic profile and the suppression level differ significantly from the static state profile. The greater the smoothing value, the more these differences are. Furthermore, the maximum suppression value at the static state amounts to 25 sigma, 21 sigma at the instantaneous velocity profile, 20 sigma at 1 s smoothing, 17 sigma at 2 s smoothing and 15 sigma at 3 s smoothing, respectively. An even greater difference is observed when comparing the profiles of the passing transport with radiation source on board. Thus, the smaller the smoothing, the greater the chance of detecting a radioactive source, but the false alarm rate also increases. This leads to the important conclusion: in ideal cases, it is desirable to choose the optimal smoothing for each vehicle of specific dimensions and average density (i.e., background suppression rate). In such case, the false alarms rate can be minimized with the optimized probability of radiation source detection. For example, for best performance, it is reasonable to set 1 s smoothing, as this profile stands most close to the profile of the instantaneous count rates. However, in order to avoid frequent false alarms, smoothing should be increased to 3 or more seconds. Since the new algorithm ensures the decision to be made at the time of the presence detector off,

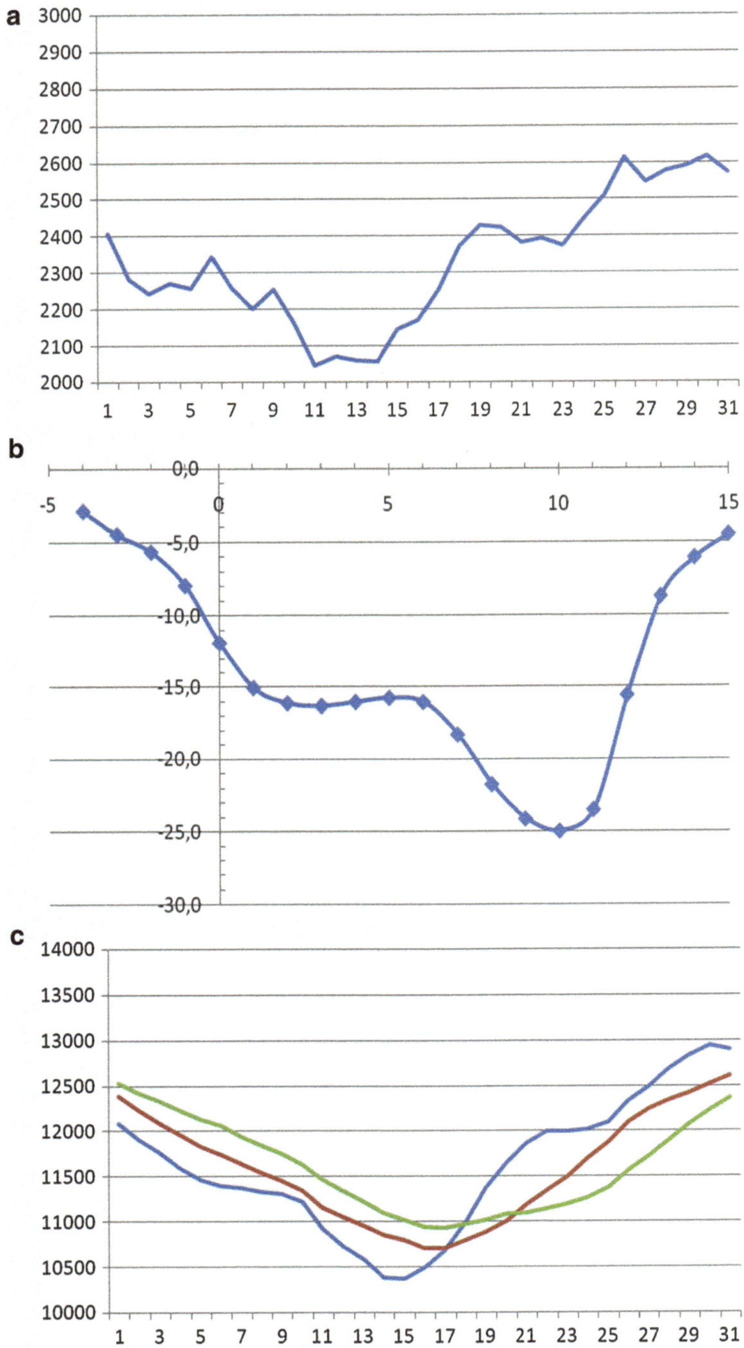

Fig. 8.3 (**a–c**) Count rate profiles for TM-850 monitor without radiation sources detected. (**a**) – instantaneous count rate profile, (**b**) – values in a steady state by step movement of transport through the monitoring control area, (**c**) – *blue line* – 1 s smoothing, *red* – 2 s and *green* – 3 s

it can automatically select the smoothing rate, depending on the type and size of the vehicle:

- passenger cars with low background suppression and short time interval between the occupancy detector on and off up to 4 s – 1 s smoothing;
- trailers with 40' shipping containers, trucks, buses, etc. with higher background suppression and time interval of 5–8 s – 2 s smoothing;
- heavy-duty trucks and trailers with high background suppression and time interval exceeding 8 s – 3 s smoothing.

Moreover, in case of doubt as to whether a radiation source is present in a passing vehicle, the algorithm also provides for parameter recalculation at various values of smoothing and analysis of the results obtained on the coincidence basis.

2. Whereas the algorithm is based on the physical principles of monitored parameter analysis, it is possible to avoid numerous test measurements using different types of vehicles without radiation sources and with major isotopes of Cobalt, Barium, and Cesium. Practically, this means that, just replacing the electronics, any monitor that uses plastic detectors can integrate the algorithm by carrying out some simple tests by the manufacturer, and then the monitor can be installed at any location with the natural background being different substantially from the factory environment. This fact is extremely important. Currently, various programs (IAEA, SLD, national programs in several countries, such as Poland) are launched worldwide and thousands of monitors with plastic detectors of various types are successfully operated. Almost all of them are equipped with traditional algorithms based on comparison of the total count rate with its threshold value, expressed in sigma. Typically, this threshold is set at 5–6 sigma, which provides the false alarm rate not exceeding 1/1000, subject to meeting the requirements of different standards for the detection of radiation sources, excluding natural background suppression, and at values typical for sea level. However, in reality, these requirements can not be met in principle. Firstly, as noted above, any vehicle causes suppression of the natural background and, as a consequence, MDA increases as well (Table 8.1). Secondly, in those locations where the natural background level exceeds the sea level values, MDA can vary significantly, even without background suppression. So, for example, the count rate of a VM-250 monitor installed at sea level site is 1,600 cps (40 sigma), and at 1,600 m altitude above sea level – 4,000 cps (62 sigma). Thirdly, at many locations, vehicles undergoing radiation control contain a rather high NORM concentration, which leads to significant increase in the number of innocent alarms. And, as noted above, users of this equipment have to substantially increase the threshold settings. ASIA-New provides for upgrade of these monitors at minimum extra cost. Besides, such monitor functions like detection of radiation sources with natural background suppression and NORM verification will be similar to those of spectroscopy-based monitors. Naturally, the cost of upgrade will be much lower than the replacement of monitors with plastic detectors by new monitors with spectroscopic functions.

8.4 Some Results of the Algorithm Tests

The current version of the algorithm implements several branches of decision trees, depending on the mode of transport:

- passenger cars with low background suppression and short time interval between the occupancy detector on and off up to 4 s;
- trailers with 40' shipping containers, trucks, buses, etc. with higher background suppression and time interval of 5–8 s;
- heavy-duty trucks and trailers with high background suppression and time interval exceeding 8 s.

Also, when triggering the parameter of high-energy radiation source detection, the algorithm switches to natural source verification, based on a new operation principle.

Software necessary to test the algorithm was developed. As references, we used the results of numerous measurements carried out in cooperation with L. Kagan, T. Gregoire and R. Rogers at Rapiscan site by using two monitors – VM-250 and TM-850. Figure 8.4 demonstrates the algorithm test site layout.

Measurements were made for different types of passenger cars – 240 passages without sources and 1,145 passages with different sources, for 40' shipping container trailer – 1,155 passages without a source and 2,431 passages with different sources on board. In addition, measurements were performed with shipping containers loaded with different amounts of KCl (2–5 t) without artificial sources, as well as with various sources on board (masking). It should be noted that the information collected from both the monitors was further processed and integrated into a single file. Thus every vehicle passage through the radiation control area of both monitors meant 1 increment for VM-250 plus 1 increment for TM-850. The

Fig. 8.4 Test site layout

Table 8.2 Results of the algorithm tests at passenger cars of various types

	"Old" ASIA			"New" ASIA		
	Passages	Alarms	Alarm type	Passages	Alarms	Alarm type
No source	240	0	0	240	0	0
Co-57	360	158	158 low energy 1	360	289	289 low energy 1
			202 no source			71 no source
Ba-133	424	373	373 low energy 2	424	424	Low energy 2
			51 no source			
Cs-137	120	99	99 mid energy	120	120	Mid energy
			21 no source			
Co-60 (small)	120	0	No source	120	120	Small Co-60
Th-232	120	5	5 NORM	120	35	35 low energy-2, mid energy, 95 no source
			115 no source			

Table 8.3 Results of the algorithm tests at trailer

	"Old" ASIA			"New" ASIA		
	Passages	Alarms	Alarm type	Passages	Alarms	Alarm type
No source	1,155	3	Low energy	1,155	1	1 low energy
Co-57	920	655	655 low energy 1	920	788	788 low energy 1
			265 no source			132 no source
Ba-133	759	650	650 low energy 2	759	714	714 low energy 2
			109 no source			45 no source
Cs-137	160	111	111 mid energy	200	183	183 mid energy
			49 no source			17 no source
Co-60 small	151	0	NORM or no source	151	0	no source
Th-232	100	5	Low energy 2	100	24	Mid energy
Co-60 + Pb (1–10 cm)	120	0	NORM or no source	120	54	54 Co-60
						66 no source

following table shows the results of ASIA-New tests. For clarity, it also provides the test results for the previous version of the algorithm. It should be noted that all the studies mainly applied such artificial sources, which could not be detected by the traditional algorithm (Table 8.2).

When monitoring 40' shipping container trailers, a series of measurements with Co-60 source was also conducted for verification of NORM in addition to detection of radiation sources. Table 8.3 shows the results for the Co-60 source activity of approximately 0.5 mCi in a lead container with wall thicknesses of 1, 5 and 10 cm. Previous "Old ASIA" algorithm detected a source at 1 and 5 cm wall container as NORM, and a source placed in a 10 cm wall container was not detected by both algorithms due to significant shielding of radiation. The new algorithm provided

Table 8.4 Results of monitor tests for NORM-masked source detection

	"Old" ASIA			"New" ASIA		
	Passages	Alarms	Alarm type	Passages	Alarms	Alarm type
KCl	1,411	1,411	NORM	1,411	1,411	NORM
KCl + Co-57	160	160	NORM	160	160	109 alarms NORM
						51 alarm low energy 1
KCl + Ba-133	160	160	151 low energy 2	160	160	1 NORM
			9 NORM			159 low energy 2
KCl + Cs-137	20	20	NORM	20	20	2 alarm NORM
						18 alarms mid energy
KCl + U-ore	20	20	Low energy 2	20	20	Low energy 2
KCl + Co-60 + Pb	40	40	NORM	40	40	24 alarms Co-60
						16 NORM

100 % detection of Cobalt source in 1 cm wall container and 70 % detection in 5 cm wall container. Cobalt source outside the container was verified by both algorithms as Co-60.

Table 8.4 demonstrates the results of tests for detection of artificial radiation sources with natural source masking (e.g., 5 t KCl). These tests are preliminary, and the investigations in this domain are still in progress.

As seen from the table, the new algorithm enhances the chance of radiation source detection with high natural background suppression and substantially solves the problem of NORM verification. Given that all measurements were taken at 1,600 m altitude above sea level, where the natural background is more than three times higher than normal, the algorithm performance is expected to be even better at locations with lower natural background.

8.5 Conclusions

1. The new algorithm is based on physical principles and phenomenology. It does not require multiple test measurements of each type of monitor to generate the statistically significant thresholds for the parameters used in making decisions on the passing through vehicles.
2. The algorithm provides for automatic selection of the smoothing value, which significantly increases the chance of radioactive source detection while meeting the requirements for acceptable false alarms rate.

3. The algorithm implements the new principle for NORM verification, where Co-60 sources with activity exceeding 0.2 mCi without shielding and above 0.5 mCi in lead shielding of up to 3 cm thickness are verified as radioactive sources.
4. Preliminary studies have also demonstrated that Co-60 NORM (5 t KCl) masking makes Co-60 source of 0.5 mCi plus activity without shielding to be identified as an artificial source.
5. The algorithm and integrated electronics can be installed at monitors of any design (including handheld monitors), provided they use plastic detectors.

References

1. Kagan L, Stavrov A (2013) Outstanding breakthrough in detection of radioactive sources by portal monitors with plastic scintillation detectors. In: The 2013 international conference on applications of nuclear techniques, Crete, Greece, 23–29 June 2013
2. Kagan L, Stavrov A (2010) Methodology of testing vehicle radiation portal monitors considering field operation conditions. In: INMM 50th annual meeting (Final program), Tucson
3. Kagan L, Stavrov A (2010) New opportunities of radiation portal monitors with plastic detectors. Third European IRPA Congress, Helsinki, Finland, 14–18 June 2010
4. Kagan L, Stavrov A (2010) Border radiation control: integral system or simple combination of instruments and rules? In: 6th symposium hazards – detection and management (Final programme), Dresden, 20–24 September 2010

Chapter 9
Mobile Units for Characterization of Contaminated Sites

Thomas Streil and Veikko Oeser

Abstract As part of an environmental, accidental or terroristic action remediation plan to be applied to areas affected by past activities, accidents or terroristic attacks, characterization of the site is a mandatory step. This activity will determine the extent of the contamination, contaminants' distribution, etc. Traditionally, this activity involves the collection of different environmental samples and laboratory analysis of the relevant radio nuclides (and eventually other contaminants like heavy metals). When the results are available they are interpreted and then a decision is made. This process is normally very expensive and time consuming. In recent years many techniques have been made available for in-situ measurement that can provide reliable information on the contamination profile in radiological contaminated land. Such measurements tend to be less expensive, faster and with the aid of GPS/GIS systems decisions can be made on-site in real time. To overcome this situation we developed the DACM (Data Acquisition and Control Module) technology. Instruments based on this technology can be modified anytime by the user without special knowledge and the claiming of the manufacturer.

The DACM based offers a set of components which can be configured, parameterized and controlled with respect to the requirements on site. Typical components are signal inputs for sensors like Co_2, Methane, So_2.., control outputs for instance for pumps, magnetic valves but also complex functional blocks like spectrometers, GPS receiver, PID regulators etc. A complex sampling schedule can be created within few minutes by a graphical software interface.

One version of this system is the NucScout as a handy and robust $2'' \times 2''$(optional $3'' \times 3''$) NaI(TI) Nuclide Identifier and quantifier. With less than 2 kg including GPS and ZigBee wireless connection, if the device is operated in inaccessible or contaminated areas, he can be so calibrated by use in 1 m high from the soil, that he show direct the nuclide activity in Bq/kg. So you can get with a time resolution of 10 s and a speed of 1 m/s a local resolution of 10 m and you can detect a specific activity less than 200 Bq/kg soil activity on the surface.

T. Streil (✉) • V. Oeser
SARAD GmbH, Dresden, Germany
e-mail: streil@sarad.de

S. Apikyan, D. Diamond (eds.), *Nuclear Terrorism and National Preparedness*, NATO Science for Peace and Security Series B: Physics and Biophysics, DOI 10.1007/978-94-017-9891-4_9

The NaI(TI) detector is also used to analyze food and material probes regarding specific nuclides (e.g. Iodine, Caesium, Americium). By means of the gamma spectrum, the net activity of six user definable nuclides is automatically calculated.

A version with a aerosol sampling head with its spectroscopy filter and its silicon detector samples continuously and detects even small quantities of aerosol carried radioactivity. Both alpha and beta radiation are measured. The spectrometric analysis allows e.g. detecting Plutonium aerosols which cannot be detected by measuring gamma radiation.

Optionally, the DACM can be connected to a portable vacuum chamber with Ion-implanted Silicon detector up to 2,000 mm^2, to allow on-site analysis of mop tests and other samples under circumstances similar to those prevailing in a laboratory. The employed vacuum pumps can be connected to a 12 V source (car battery).

All detectors can be operated simultaneously. The concept of the system allows an easy handling and a standardized data basis.

9.1 Introduction

As part of an environmental remediation plan to be applied to areas affected by past activities and accidents, characterization of the site is a mandatory step. This activity will determine the extent of the contamination, contaminants' distribution, etc (see IAEA Safety Guide No. WS-G-3.1 – Remediation Process for Areas Affected by Past Activities and Accidents). Traditionally, this activity involves the collection of different environmental samples and laboratory analysis of the relevant radionuclides (and eventually other contaminants like heavy metals). When the results are available they are interpreted and then a decision is made. This process is normally very expensive and time consuming. In recent years many techniques have been made available for in-situ measurement that can provide reliable information on the contamination profile in radiologically contaminated land. Such measurements tend to be less expensive, faster and with the aid of GPS/GIS systems decisions can be made on-site in real time. Many States facing the challenge of implementing environmental remediation projects do not have adequate analytical infrastructure for site characterization. They will then need to make this available prior carrying out site characterization efforts. Therefore the place of remedial works implementation will be reduced or these activities will not be implemented at all. Mobile units may also be useful to member states who do have laboratory analysis facilities, but are faced with large, unforeseen characterization challenge, such as following and accident or radiation emergency.

9.2 DACM System Technology

SARAD's DACM (Data Acquisition and Control Module) technology offers a universal platform for instrumentation which can be fitted to any application without changes in the hardware or special knowledge in computer programming. Compared

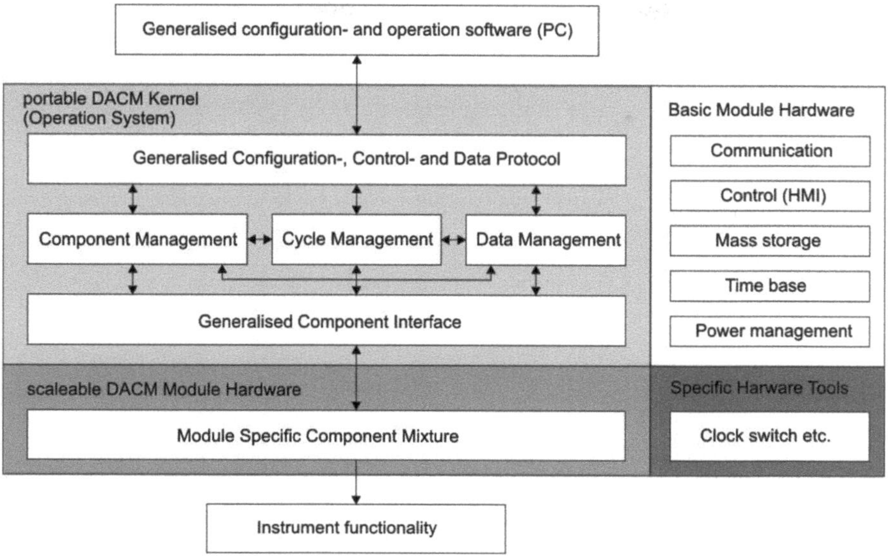

Fig. 9.1 SARAD's DACM (Data Acquisition and Control Module) topology

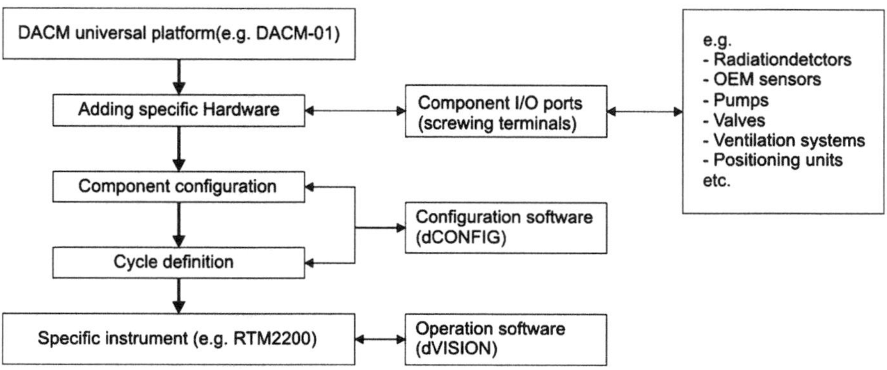

Fig. 9.2 DACM instrument synthese

with conventional data acquisition systems, the DACM is able to control complex sampling procedures in addition to the data logging and management. Scalability and portability make the DACM architecture to a universal and future-oriented system. The system performance can be expanded or specialised in any order while the basic structure of system control and data flow is not touched. The user gets the possibility to modify afterwards all instrument functions with respect to the own requirements. Each DACM based instrument contains a system core and a set of so-called components (see Figs. 9.1, 9.2, and 9.3). The system core includes the basic instrument properties like device control, data storage, interfacing, component

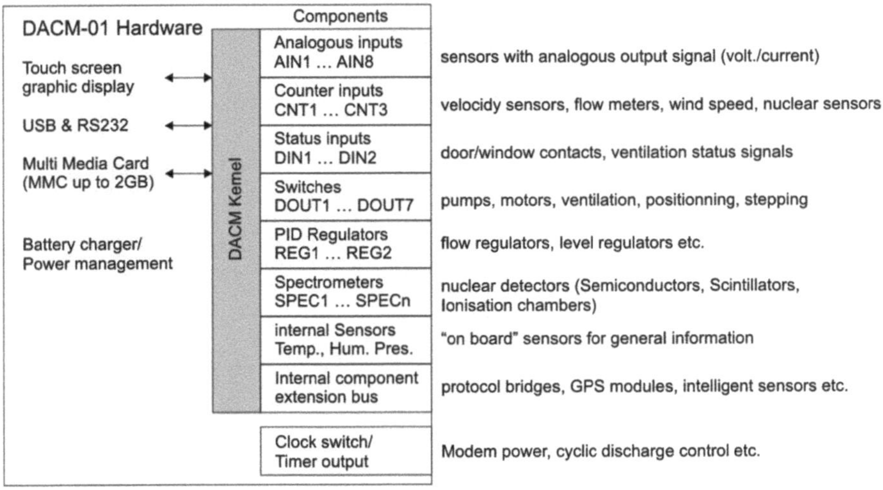

Fig. 9.3 DACM hardware basic implementation

management and the sampling control. The components implemented in the DACM can be considered as a library of functional units which can be manipulated to fit them to own requirements:

- Each component can be configured specifically (e.g. transfer function, operation mode)
- The state of a component can be changed during the sampling cycle
- components can generate, display and store measurement data
- components can generate alerts or react to alert states.

9.2.1 Sampling Control

To set and change the component status during the data acquisition process, at first a so-called sampling cycle has to be defined. The cycle can be also interpreted as the basic storage interval because each measurement series consists of a number of repetitions of this cycle. All acquired data are stored at the end of the cycle even if the physical sampling has been taken at any time within this cycle.

A cycle can be split into a maximum of 32 sequential subdivisions. The status of each component can be defined for each subdivision by the user. This is realised by a control bit or a control word. The time period of a subdivision can be set from one second up to the cycle interval period in one second steps.

The configuration software running on a PC allows editing the cycle definitions easily. All components are listed in a table and their status information is shown as a time chart. Only a simple mouse click into the time chart is necessary to change the status of any component within a subdivision. Several cycles may be stored on the DACM's non-volatile memory and called for execution.

9.2.2 *Component Configuration*

Each implemented component can be separately configured by the user. Number and type of the configuration parameters are defined by the type of the component. A module can contain a maximum of 64 components. These components could be of the same or different types.

9.2.3 *Data Flow*

All generated data is stored in binary format as it has been generated by the component of origin. This ensures a 100 % re-traceability of the acquired data resulting in a high level quality assurance. A component can generate one or more measurement values from the basic data. To display and transmit recent readings, all results are calculated by the module using these basic data. If the complete data is transferred to the PC, the packed basic data are transmitted together with all necessary module and configuration information.

That means, the calculation procedures are implemented in the DACM as well as in the PC software.

If the data will be stored on the replaceable card memory (SD card), a copy of the whole module and component configuration will be saved in parallel on the card. Thus, the results can be calculated correctly independent on the origin of the data.

9.2.4 *Alert Management*

Several components are able to generate alerts (alert source). The type of alert is defined by component type and required alert conditions (alert level etc.). Both are elements of the component configuration. The alert check is carried out once per Second for all components which are enabled and activated during the recent cycle subdivision.

Other components are able to process alerts, meaning to act as an alert destination. If a component acts as alert source, a complete list of all available alert destinations is provided to select the component which shall react in case of an arising alert.

How the component responds to the alert situation is also defined by the component type and configuration of the alert destination.

9.2.5 PC Interfacing

Operation and configuration software are delivered as two separate programs. Thus, once configured, the instrument looks like an instrument designed only for that purpose. This simplifies the handling and operation of the unit if it is used by unskilled staff.

The configuration software allows to:

- transfer the configuration information and the cycle definitions from and to the module
- edit the common module information
- edit the configuration of the implemented components
- edit the cycle definitions
- save the configuration and cycle information on the hard disc.

After loading the actual module and component configuration from the DACM or hard disk, a list of all components available in DACM will appear. Several tables show clearly arranged the settings of the configuration parameters for all components of the same type.

A simple mouse click into the table opens a dialog window to edit the parameter settings of the selected component. The changed parameters can be transferred to the DACM ore saved on the hard disk after finishing the configuration process.

The cycle definition is also very easily editable by a time chart (cycle subdivisions in X axis) which lists all components vertically. Subdivisions can be created or deleted and component states can be changed by a few mouse clicks clicking onto the desired chart position.

9.2.5.1 Operation Software (dVISION)

The operation software allows to:

- set the DACM real time clock
- set the clock switch
- start and stop the cycles uploaded to the module
- display and transmit the recent measurement results
- transmit the complete measurement data
- save measurement data as set or as single tracks
- show the measurement data in configurable diagrams
- export the measurement data to Excel/GIS compatible test files.

The data loaded from the module will be saved as binary data file with predefined filename and folder structure. This file contains all measurement data of the components, the configuration of the components and the module information.

For further data processing this file is split into several data tracks, one for each measurement value. These tracks can be saved and exported separately. Any combination of tracks can be loaded into the chart view later on. This allows the simultaneous displaying of data generated by several instruments or acquired during several time periods.

Graphic options:

- Selection of any time period for the chart view
- Combination of various data tracks, each with manually or automatically scaled Y-axis
- Selecting line width, line colour and line style of any data track
- Grid, cross line cursor and sliding result box for a selected track
- Scale definition for each track.

9.3 Experimental Results

The objective of the mission was to assess the efficacy of the remediation actions undertaken in a site that were contaminated in the past with NORM (high Radium content in charcoal used as absorber for Iodine production) in regard to surface radiological hazards. A remediated area was analyzed by mobile devices. Mobile Unit demonstration activities were performed in the vicinity of Baku, Azerbaijan [1].

9.3.1 Nuclide Specific Gamma Spectrometry

The NucScout (see Fig. 9.4) is a handy and robust 2″ × 2″ (optional 3″ × 3″) NaI(TI) Nuclide Identifier and quantifier. With less than 2 kg including GPS and ZigBee wireless connection he can be so calibrated by use in 1 m high from the soil, that he shows direct the nuclide activity in Bq/kg from up to 6 nuclides, which can be chosen from a big nuclide library with more than 50 nuclides. So you can get with a time resolution of 10 s and a speed of 1 m/s a local resolution of 10 m and you can detect a specific activity less than 200 Bq/kg soil activity on the surface (see Fig. 9.5).

The measured value showed that in this case the 200 Bq/kg was only a little bit higher in areas which were till now not remediated (Fig. 9.6).

In Figs. 9.7 and 9.8 is shown like with a A2M 4000 area monitor in a short time can be measured several data. The comparison of the Bi214 count rate and Radon soil gas concentration s in a good agreement. The data show also that in covering material (after remediation process) the Thoron concentration is a little bit higher because of natural small increased Thorium concentration in this soil material. But the dose rate of 200 nSv/h was in the examined area not extended.

Fig. 9.4 NucScout with wireless 2.4 GHz ZigBee data transfer

Fig. 9.5 NucScout screening measurement of Bi214 (Bq/kg) soil activity from a location in Azerbaijan

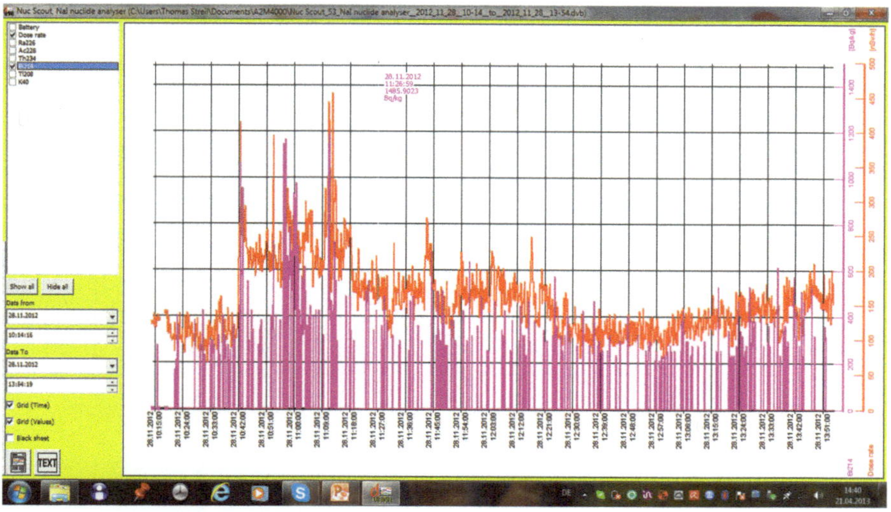

Fig. 9.6 Show a slightly increased surface soil activity in a deposit for the contaminated material, which was not fully covered by protective soil layers

Fig. 9.7 A^2M4000 monitor for measuring the Radon/Thoron soil gas activity and the local Gamma activity

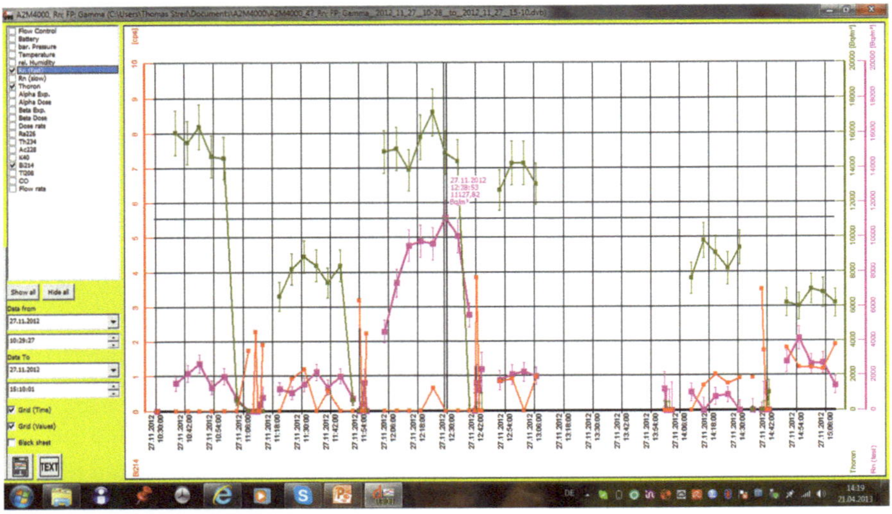

Fig. 9.8 Radon/Thoron soil gas activity and the Bi214 count rate of the screened area in Azerbaijan

9.4 Further Applications of the Mobile System

Definition of a Local Dose, Detection of Radioactive Sources The handy and robust 2″ × 2″ NaI(TI) detector is connected to the unit via a 10 m long cable, so that it can be positioned flexibly in relation to the source. Thanks to the big detector volume, even small sources can be detected. Further direct measurements in boreholes are possible.

9.4.1 Bio/Geo Scout

Net Activity of Free Definable Nuclides in Food and Material Probes The NaI(TI) detector is also used to analyze food and material probes regarding specific nuclides (e.g. Iodine, Caesium, Americium). By means of the gamma spectrum, the net activity of six user definable nuclides is automatically calculated. The Bio/Geo-Scout (see Fig. 9.9) allows the fast and reliable detection of radioactivity in food and soil samples. A large volume (2″× 2″) NaI detector (used in Marinelli-geometry) in combination with a Gamma spectroscope gives the possibility to detect six pre-defined nuclides like Americium, Cesium or Iodine. The results are presented in Bq/kg by measurement of the sample weight using the internal scale.

In order to minimize the required sample time, the instrument should be placed at locations with low background radiation. Increased background (>100 nSv/h)

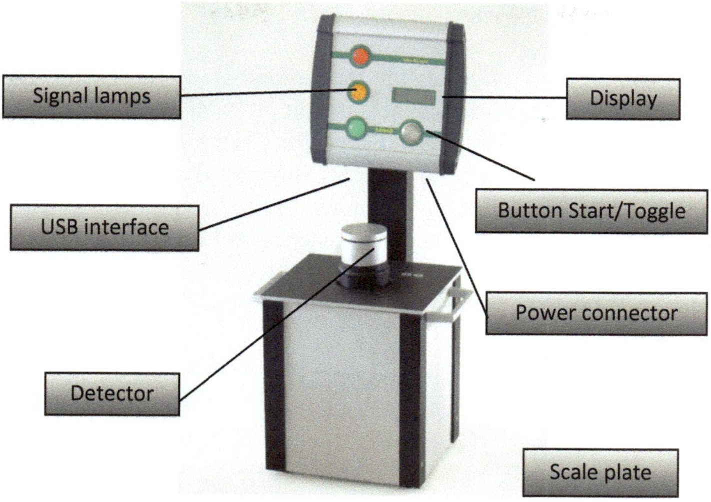

Fig. 9.9 Show the new mobile device Bio/Geo Scout

can appear in buildings with walls/floors made from natural stones (Granite, Marble etc.) or inside rooms with high Radon concentrations. If the unit shall be used in radiological laboratories, other radioactive sources must not be stored close to the Bio-Scout.

9.4.1.1 Setting the Bio-Scout in Operation

Before connecting the power adapter, an empty sample container (Marinelli beaker) must be placed at the sampling desk. This is required for the zero-calibration of the internal scale. As soon as the power is present, all three lamps will light for a short time and the buzzer gives an acoustic signal. A 10 min measurement starts automatically to determine the local radiation background. If the instrument has been relocated, this procedure must be repeated (disconnect/connect power adapter). The display shows the remaining time for background measurement. After that appears "Press START for first sample".

9.4.1.2 Filling the Sample into the Beaker

It is strictly required to fill the food sample correctly into the Marinelli beaker. The beaker must be always filled completely with 1 l sample volume. This volume gives the reference fort the density calculation at the one hand and ensures a correct sampling geometry at the other hand. If sampling beakers with geometries different from the original shall be used, a re-calibration of the unit is required.

9.4.1.3 Carrying Out a Measurement

The analysis can be started by pressing the button below the display. The display shows after weighting the sample net weight as well as the required sample time to achieve the detection limits. For a 1 kg sample and a background less than 100 nSv/h, a sample interval of approximately 60 s is expectable. Higher background or less weight results in longer intervals. If after this period some significant activity has been detected, the analysis will be continued automatically for the same period to decrease the statistical fluctuations. The end of each measurement is indicated by an acoustic signal and lighting up of one of the lamps.

9.4.1.4 Green Lamp and Short Signal

No activity in the range of the adjusted limits has been detected. The display shows "BELOW LIMIT Press START for next sample".

9.4.1.5 Red Lamp and Interrupted Signal

An activity above the limit has been detected at least for one of the nuclides. A list of detected nuclides appears on the display. Push the button to toggle the display to show the results for each nuclide. The button must be pressed for a longer period (until second beep) to go back to the measurement mode. To show the results again, press.

9.4.1.6 Yellow Lamp and Interrupted Signal

An activity above 50 % of the alert limit has been detected at least for one of the nuclides.

The display function is the same like for red indicator. It is recommended to repeat the measurement.

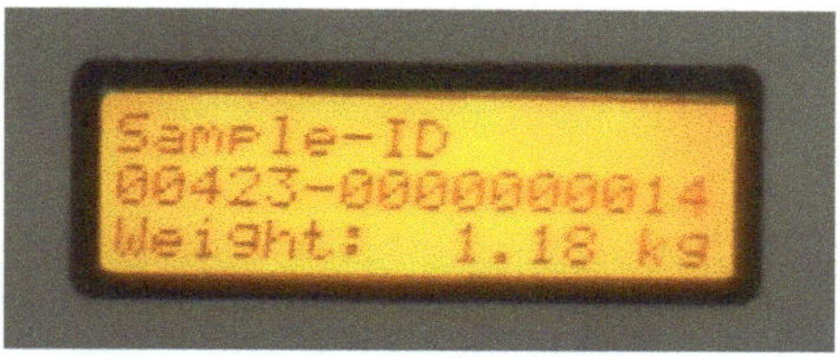

9.4.1.7 Predefined Nuclides, Detection Limits and Alert Limits

In case of nuclear accidence, a number of nuclides is used as tracers such as I-131, Sr-90, Pu-239 und Cs-134/137. Additionally, large amounts of Te-132 and Ru-103/106 were emitted after reactor crash. Because of their half life times and mobility, especially Cesium, Strontium and Plutonium are of interest for food samples. Strontium and Plutonium are not or very difficult to detect by gamma spectroscopy while Cesium is good detectable even with NaI detectors. Table 9.1 shows the list of the six selected nuclides detected by the Bio-Scout.

The analysis procedure presumes that the sample does not contain any activity from natural occurring nuclides except K-40 (Potassium). This is the case for any food sample. The Bio-Scout is not suitable to measure soil or material samples containing elements of the Uranium or Thorium decay chain. In this case, the activities are interpreted as an increase of the natural background leading to a new background measurement. Many foods contain Potassium so that the unit tolerates K-40 up to a level of 200 Bq/kg.

Table 9.1 Selected nuclides with detection and alert limits

Nuclide	Detection limit [Bq/kg]	Alert limit Bq[kg]
I-131	100	200
Cs-137	100	100
Cs-134	100	100
Ru-103	100	–
Te-132	100	–
K-40	100	200

9.4.1.8 USB Interface

The acquired gamma spectrum of a sample will be stored in case of alert (up to 70 spectra). The USB interface allows the download of the spectra for further assessment. Furthermore, all settings with regarding the nuclide definition and analysis parameters are done via this interface. The receptacle is placed on the bottom of the control panel (left side).

9.4.2 Experimental Results

The following documentation include brief characteristic of the SARAD new Bio/Geo-Scout measurement device as also test results of comparison to the High-purity germanium detector performed in Chiba, Japan (Fig. 9.10) [2] (Table 9.2). With samples from the Fukushima accident with the origin between Chiba and Fukushima like Fig. 9.11 show (Fig. 9.12).

Table 9.3 show that, if a quick and a precise radiological inspection of biological or geological samples is required Bio Scout is a very appropriate, cheap and last but not least easy to use alternative for more advanced scientific solutions.

9.4.3 Measurement of Radioactive Aerosols in Inhaled Air

The aerosol sampling head with its spectroscopy filter and its silicon detector samples continuously and detects even small quantities of aerosol carried radioactivity. Both alpha and beta radiation are measured. The spectrometric analysis allows e.g. detecting Plutonium aerosols which cannot be detected by measuring gamma radiation. The Data transmission and device control can be done by GPRS or GSM modems, as well as via ZigBee adapter (Wi-Fi), if the device is operated in inaccessible or contaminated areas.

Fig. 9.10 High-purity germanium detector, NIRS, Chiba

Table 9.2 Comparison of detectors selected features

PARAM/DEVICE		Standard HPGe	Bio/Geo Scout
Weight		>500 kg (With shield)	6 kg (No shield required)
Dimension		$>1.5 \times 0.5 \times 0.5$	$=0.5 \times 0.25 \times 0.25$
Portability		No (Stationary only)	Yes (Mobile)
Price	>100 k€	~5 k€	

9.4.4 Mop Tests, Surface Contamination (Clothes), Electrochemical Probes

Optionally, the DACM can be connected to a portable vacuum chamber with Ion-implanted Silicon detector up to 2,000 mm^2, to allow on-site analysis of mop tests and other samples under circumstances similar to those prevailing in a laboratory. The employed vacuum pumps can be connected to a 12 V source (car battery). All detectors can be operated simultaneously. The concept of the system allows an easy handling and a standardized data basis. The device offers predefined measurement procedures that can be easily modified by the user. Additional measurement programs can be created without any problem.

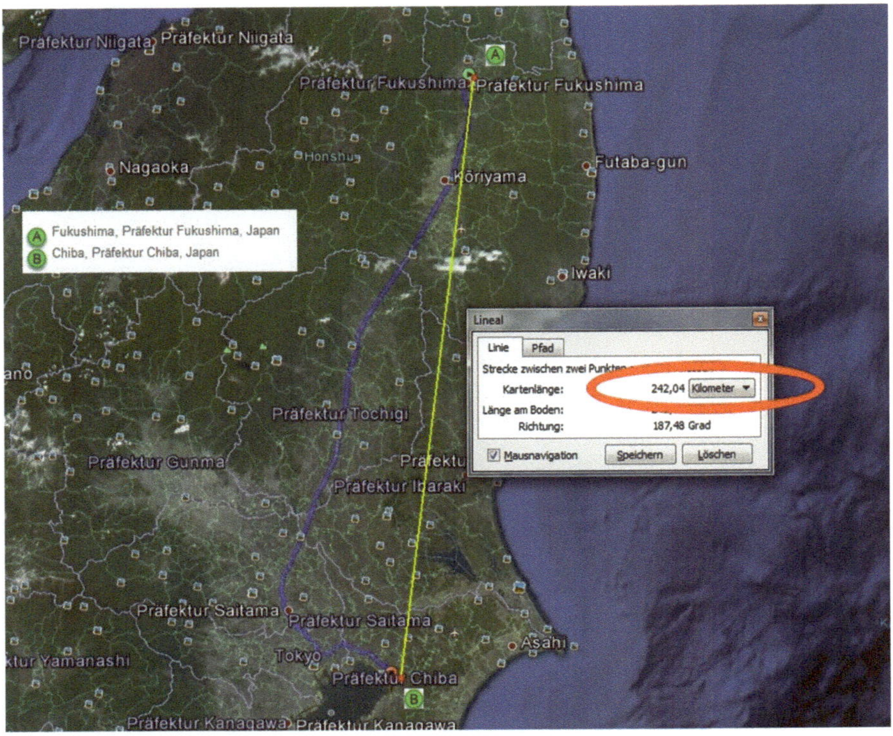

Fig. 9.11 Chiba-Fukushima linear distance: 242 km location of the test measurements in respect to Fukushima

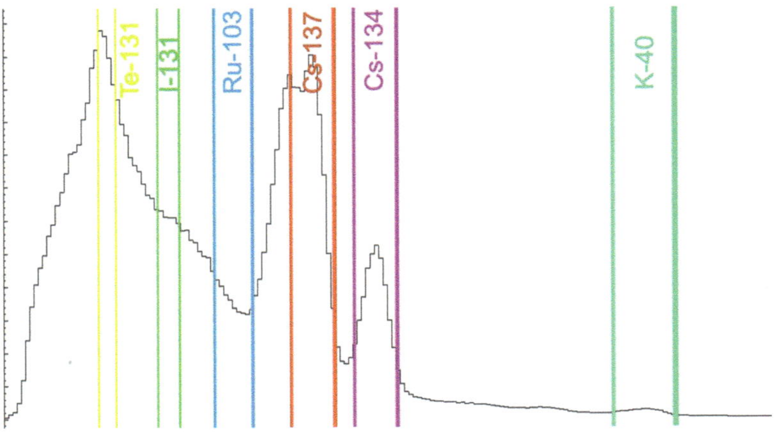

Fig. 9.12 Sample spectrum – Fukushima, 0,60 [kg], 488 [sec]

Table 9.3 Comparison of measurement results between bio (Geoscout and HPGe Det)

SARAD Bio Scout				
Sample name: NIRS-soil-wet, Location: Chiba	Weight [kg]	**0.94**	Interval [sec]	**206**
Measurement parameters	Weight [kg]	**0.94**	Interval [sec]	**206**
Nuclide [Bq/kg]	Cs-134	±Δ[%]	Cs-137	±Δ[%]
1.	**150**	**25**	**230**	**25**

HPGe (High-purity germanium detector)				
Sample name: NIRS-soil-wet, Location: Chiba				
Measurement para parameters	Weight [kg]	**0,078**	Interval [sec]	**600**
Nuclide [Bq/kg]	Cs-134	±Δ[%]	Cs-137	±Δ[%]
3.	**145**	**9**	**279**	**6**

9.5 Conclusion

The used mobile devices for site Characterization in environmental remediation projects from NORM material (high Radium content in charcoal used as absorber for Iodine production) show the possibility in less than 2 days to examine an area of more than 30 ha. Such mobile systems are also possible to use for the characterization of the contamination after terroristic attacks (dirty Bombs) or nuclear accidents like Fukushima. The new DACM technology allows creating device with special customer specifications. This is a big step to reduce the time a cost consuming lab procedures and show in time the results.

References

1. IAEA report (2012) Results of surface radiological characterization of past NORM contaminated sites of Ramana and Surakhany, 26–30 November 2012, Azerbaijan
2. Private communication Miroslaw Janik, Ph.D. National Institute of Radiological Sciences (NIRS), 4-9-1 Anagawa, Inage-ku, 263–8555 Chiba, Japan, e-mail: mirek@fml.nirs.go.jp

Chapter 10
Real-Time Radionuclide Identification and Mapping Capabilities of the U.S. Environmental Protection Agency's Airborne Spectral Photometric Environmental Collection Technology

John Cardarelli II, Mark Thomas, Timothy Curry, Robert Kroutil, Jeff Stapleton, David Miller, Gary Small, and Brian Dess

Abstract The U.S. Environmental Protection Agency, CBRN Consequence Management Advisory Team fields a fixed-wing aircraft known as the Airborne Spectral Photometric Environmental Collection Technology (ASPECT). ASPECT is a 24/7/365 response-ready asset that can be airborne within an hour and collecting chemical, radiological and photographic data anywhere in the continental United States within 9 hours of notification from its home base near Dallas, TX. A primary goal of the program is to provide actionable intelligence to decision makers within minutes of data collection via the aircraft satellite communication system while the aircraft is still flying. To achieve this goal, a new method to process airborne gamma spectroscopy data was investigated that effectively eliminates the need for traditional airborne calibration methods currently used in airborne systems (e.g., stripping coefficients, test lines, radon correction, altitude, and background

---

J. Cardarelli II (✉)
CBRN Consequence Management Advisory Division, Environmental Protection Agency, Erlanger, KY, USA
e-mail: Cardarelli.John@epa.gov

M. Thomas
CBRN Consequence Management Advisory Division, Environmental Protection Agency, Kansas City, KS, USA

T. Curry
CBRN Consequence Management Advisory Division, Environmental Protection Agency, Kansas City, KS, USA

R. Kroutil • J. Stapleton • D. Miller
Kalman and Company, Inc., Virginia Beach, VA, USA

G. Small • B. Dess
Department of Chemistry & Optical Science and Technology Center, University of Iowa, Iowa City, IA, USA

S. Apikyan, D. Diamond (eds.), *Nuclear Terrorism and National Preparedness*, NATO Science for Peace and Security Series B: Physics and Biophysics,
DOI 10.1007/978-94-017-9891-4_10

corrections). The new algorithm provides nuclide identification near real-time and creates an radionuclide-specific map without user-interface input.

This algorithm uses a novel combination of signal processing and multivariate pattern recognition to implement an automated classifier for the detection of radionuclides from gamma-ray spectra. The algorithm was applied to detect Cs-137 over fallout from nuclear weapon tests conducted in the 1960s. The aerial survey data were collected in 2010, over 50 years post-detonation. The combination of infinite-impulse response (IIR) filtering with piecewise linear discriminant analysis (PLDA) classification was found to yield the most sensitive detection method.

Overall, this method has demonstrated very promising results and appears to be generalizable to other aircraft configurations provided they fly within an acceptable altitude range and are equipped with sodium iodide crystals.

10.1 Introduction

The US Environmental Protection Agency Airborne Spectral Photometric Environmental Collection Technology (ASPECT) Program provides assistance to the first responder by providing an aerial tool to remotely detect chemicals and radiation, take photos (infrared and visible), and relay this information directly to decision makers in the field [1]. Since 2001, ASPECT has assisted the response community in over 140 deployments ranging from lost neutron sources to pre-deployments to support Preventive Radiological Nuclear Detection (PRND) missions in Homeland Security applications. The aircraft is located near Dallas, Texas and is airborne within 1 h of activation. It is equipped with (1) a high speed Fourier transform infrared spectrometer (FTIR) coupled with a wide-area infrared (IR) line scanner for chemical detection, (2) IR video and digital camera for imaging and situational awareness, (3) 25 l of sodium iodide (NaI) and 1 l of lanthanum bromide detectors for gamma-ray detections, (4) 2 Boron Trifluoride (BF3) neutron detectors, and (5) an onboard processing system coupled with a high speed satellite communication system to process and transmit preliminary results to a reachback team for quality assurance before being delivered to the decision makers.

Aerial gamma spectroscopy has a distinct set of protocols and challenges [3, 4]. A typical survey follows a fairly complex protocol requiring a system of properly functioning instrumentation (e.g., aircraft, "gain stabilized" radiation detectors, radar altimeter, GPS) and defined flight parameters (e.g., cosmic and background flight lines/areas, target survey areas, line spacing, speed, and height). It often involves days or weeks of preparation and analysis. Commonly the aircraft is calibrated prior to the survey to obtain sensitivity coefficients for various radionuclides and to account for crosstalk among radionuclides (stripping coefficients) and radiation contributions from the aircraft, radon, and cosmic radiation. Generally an aerial survey involves flying background surveys while accounting for height differences, changes in cosmic radiation with altitude, and air attenuation based on humidity, temperature, and pressure.

Background suppressed pattern recognition (PR) is a method used by the ASPECT Program to process infrared spectroscopy to detect selected chemical compounds [7]. Traditional methods of IR spectroscopy require that each spectrum potentially containing an unknown chemical be referenced against a standard background spectrum. The ratio of the two spectra effectively removes instrumental and background-related features, thereby permitting the ratioed spectrum to be compared against a reference library of known chemical signatures. For this method to be effective, the background environment must be consistent between the two samples; otherwise, spectral artifacts will be introduced. This requirement is complicated and not often satisfied in an airborne setting since the spectral background is constantly changing. Work by Small et al. [6] has shown that infrared-active chemical species exhibit a unique pattern within the FTIR interferogram, permitting data analysis to be directly performed within the interferogram space of the data. Background information is suppressed by using unique time-dependent digital bandpass filters to "window" a portion of the interferogram, thus providing a pattern to be submitted to multivariate classification techniques such as piecewise linear discriminant analysis (PLDA) [5]. Given a set of representative training data, the output of the PLDA algorithm is a classification model that can be applied to future data to obtain a yes/no decision regarding the presence or absence of the target chemical that serves as the basis for the model.

This technique has been found to provide robust detection of weak chemical signatures within the complex IR radiometric environment. Specifically, once adequately trained for a given compound, the method can detect signatures which often cannot be observed in single spectra using traditional background subtraction methods. This detection sensitivity has been demonstrated in both laboratory and operational settings, and has been achieved with a negligible rate of false detections. Finally, the background suppressed pattern recognition method is well suited for automation, a crucial attribute during emergency response missions.

10.2 Background Suppression Pattern Recognition Applied to Gamma Spectroscopy

In the work described here, the background suppressed PR methodology described above has been applied to gamma spectroscopy. The method is based on the use of a set of 25,000 background spectra collected throughout the United States that characterize gamma radiation spectral variations due to different altitudes (up to 350 m), and variations in natural background and radon concentrations. This dataset essentially eliminates the need to conduct (1) calibration flights over land and water, (2) background flights, (3) aircraft-specific stripping coefficient corrections, (4) radon corrections, and (5) altitude adjustments. This background spectral database is used in preprocessing collected gamma spectra to suppress background features and in the assembly of the training set needed to derive the radionuclide-specific coefficients that comprise the PLDA classification model. Suppressing the background signature from the airborne spectral data results in an radionuclide-specific identification that can be detected and mapped in an automated manner in near real-time. This method has the potential to: reduce operational and maintenance costs of having to routinely calibrate an aircraft over calibration fields, produce near real-time radionuclide-specific products with confidence intervals, eliminate potential bias by automating the analyses in an objective process, and be applied to different aircraft platforms (e.g., helicopter and fixed wings) without the need to collect aircraft-specific calibration data.

10.3 Pattern Recognition Methods

Given an input gamma spectrum containing n spectral channels, the PR method employed in this work was based on the following steps: (1) use of the database of background spectra described above to fit and subsequently remove a first approximation of the background components in the acquired spectrum, (2) application of an infinite impulse response (IIR) digital filter to the obtained residual spectrum to provide further suppression of noise and background features, (3) isolation of a contiguous segment of p spectral channels within the filtered residual spectrum, (4) submission of the resulting p-dimensional spectral "pattern" to a previously computed PLDA classification model, and (5) mapping of the discriminant score obtained from the PLDA model onto a classification probability scale that can be used to assess whether the radionuclide targeted by the model is present or absent.

To facilitate the background subtraction, the background database was sorted by altitude into 15 m bins, and the average background spectrum was computed within each bin. For the input spectrum whose classification was sought, its associated radar altimeter reading was used to retrieve the average background spectrum from the corresponding altitude bin. Linear regression was then used to fit the input spectrum to the retrieved background, and the resulting residual spectrum was obtained.

For a given target radionuclide, implementation of the algorithm steps described above required optimization of the following: (1) the frequency response of the IIR digital filter to be applied to the residual spectrum obtained from the background subtraction step, (2) the starting and stopping spectral channels used to define the pattern that served as the input to PLDA, (3) the coefficients that comprised the PLDA classification model, and (4) the function used to convert the PLDA discriminant score to an associated classification probability. The filter parameters (stopband and passband edges, degree of stopband attenuation) were initially optimized by a manual procedure in which field-collected spectra with (i.e., active spectra) and without (i.e., inactive spectra) the signature of the target radionuclide were filtered and then processed by principal component analysis to obtain a visual display of the degree of discrimination between the active and inactive spectra.

Candidate filters were then incorporated into the optimization of the remaining parameters.

A grid search procedure was used to identify the best combination of filter bandpass and spectral segment. For the Cs-137 classifier, filtered spectral segments were searched over the range of 414–713 keV in steps of 3 keV (i.e., one spectral channel) for segment widths of 354, 384, 414, and 444 keV. These segment widths corresponded to 120, 130, 140, and 150 contiguous spectral channels and focused on the characteristic gamma-ray emission at 662 keV. The classifier for Co-60 searched segments of the same widths, but the starting point was varied over 771–1,428 keV in steps of 6 keV (two spectral channels). This search focused on the characteristic emissions at 1,173 and 1,332 keV.

For each combination of filter and spectral segment, a training set of 15,000 patterns (3,000 active, 12,000 inactive) was used to optimize a PLDA classification model that provided the best discrimination between the active and inactive patterns. The inactive spectra in the training set were drawn from the background spectral database, while the active spectra were obtained by mathematically superimposing the known spectral signatures of Cs-137 or Co-60 onto randomly selected inactive spectra drawn from the database of spectral backgrounds. The active signatures were randomly scaled such that a distribution of signal intensities was obtained that achieved approximately 85 % separation of the active data class with the top-performing classification models. This ensured that the data space formed from the active and inactive patterns had enough overlap of the data classes such that the classification model was able to achieve the most sensitive detection possible.

The top performing classification models obtained from the PLDA training procedure and grid search optimization were subsequently applied to a monitoring set of similarly constructed independent test data (i.e., 5,000 synthesized active patterns and 5,000 inactive background patterns) in order to identify the top classifiers for use in testing with field-collected spectra containing known signatures of Cs-137 and Co-60. Results are presented below for two selected classifiers. The Cs-137 model employed a bandpass digital filter with passband edges at 0.01 and 0.03 (normalized frequency) and 20–30 dB of stopband attenuation. The spectral segment was 444 keV in width, starting at 590 keV. This classifier achieved 90.8 % correct classification of the active class in training with no false detections. With the monitoring set, the rates of missed and false detections were 6.4 % and 3.7 %, respectively. For the Co-60 classifier, a lowpass IIR filter was used with the upper limit of the passband at 0.05 and a stopband attenuation of 60 dB. The spectral segment was 444 keV in width, starting at 1,156 keV. This classifier achieved 96.7 % correct classification of the active class in training with no false detections. With the monitoring set, the rates of missed and false detections were 10.7 % and 1.2 % respectively.

Classification models based on PLDA produce values termed discriminant scores as their output. Discriminant scores greater than 0.0 signal classification into the active class while negative discriminant scores correspond to inactive patterns. For the Cs-137 classifier, sufficient active signatures were present in the Sedan survey data (see below) to allow assignment of classification probabilities to the discriminant scores. This assignment took the form of a bi-Gaussian function fit to a reference distribution of estimated correct classification percentages as a function of discriminant score. The correct classification percentages were determined by visual inspection of spectra to discern the presence of the characteristic 662 keV emission. The fitted bi-Gaussian function was evaluated at 70, 95, and 99 % probabilities to determine critical values of the discriminant scores corresponding to these confidence levels.

10.4 Results and Discussion

In October 2010, the ASPECT aircraft completed a series of environmental characterization surveys at the Nevada Nuclear Security Site (NNSS), formerly known as the Nevada Test Site. The objectives of these surveys were to (1) create a cooperative program between EPA and the National Nuclear Security Agency, (2) leverage the nation's limited inventory of calibrated aerial systems, and (3) strengthen response capabilities to ensure efficient airborne measurement, identification, and assessment of radiological deposition following a nuclear/radiological incident.

Following these surveys, the ASPECT Program initiated a research project to investigate if the PR methods developed previously for chemical detection could be applied to the detection of ionizing radiation. The initial focus was to apply PR methods to the gamma-spectral data to develop radionuclide-specific (Cs-137 and Co-60) coefficients. Classification models were developed as described above and subsequently applied to the spectra acquired during three aerial surveys. None of these survey data were used in the development of the classification models. As noted previously, visual inspections of spectra collected during the Sedan survey were used in the assignment of classification probabilities to the discriminant scores output by the Cs-137 classifier.

The Sedan test was a shallow underground nuclear test (104 kT) conducted in July 1962 as part of Operation Plowshare, a program to investigate the use of nuclear weapons for civilian purposes. Results from the aerial survey show a high correlation between the traditional spectroscopy methods (e.g., region of interest; ROI) and the new PR classification model for Cs-137 (Fig. 10.1).

The PR method was further evaluated by collecting aerial gamma spectrometry data over two sources (925 MBq Cs-137 and 185 MBq Co-60) at various altitudes and offsets at the Desert Rock Airport on the NNSS. Figure 10.2 shows traditional contoured images based on total counts (30 keV to 2,811 keV), Cs-137 (600 keV to 720 keV), and Co-60 (1,092 keV to 1,416 keV) ROIs from aerial surveys conducted at 30.5 m above ground level (AGL) with 30.5 m line spacing. The PR method applied to these data eliminated the false-positive detection due to excess Compton scatter from Co-60 emissions into the Cs-137 ROI when compared to the traditional method. The PR method also shows positive detections at greater than 99 % confidence at the outermost flight lines. These results demonstrate the specificity and potential increase of sensitivity of the PR method compared to

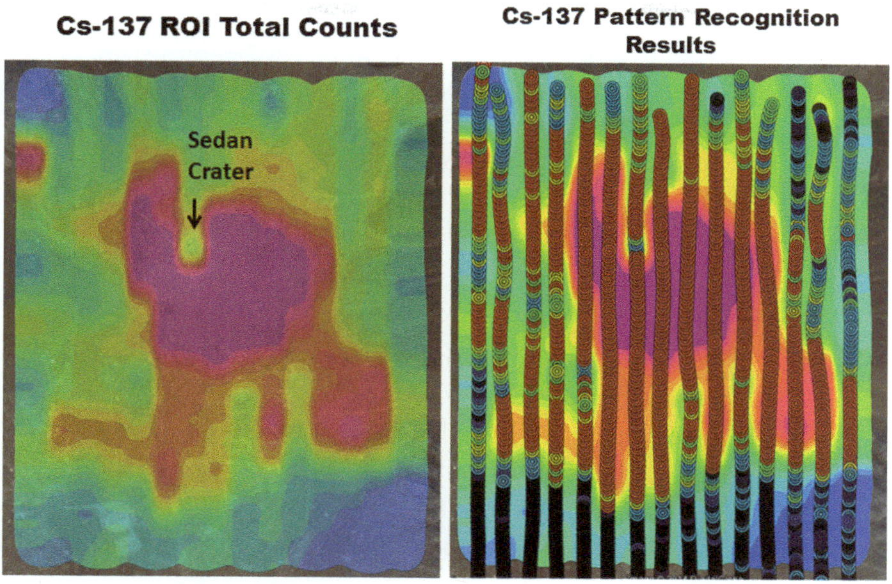

Fig. 10.1 Correlation between Cs-137 traditional gamma-spectrometry methods (*contour* on *left*) and pattern recognition results overlaid on the *contour* (*colored dots on right*). *Color codes* are (1) *red* – > 99 % confidence Cs-137 is present; (2) *yellow* – > 95 %; (3) *green* – > 70 %; (4) *light-blue* – positive classification for Cs-137 at < 70 % confidence; (5) *dark blue* – negative classification at < 70 %; (6) *purple* -70–95 % confidence Cs-137 is not present, and (7) *black* – > 95 % confidence Cs-137 is not present

traditional methods and the lack-of-need for determining stripping coefficients for PRND or environmental survey missions.

The Cs-137 PR coefficients were further tested by use of data collected over the fallout from the Small Boy nuclear test. An area of about 6 mile^2 downwind from the detonation location was surveyed by the ASPECT aircraft (20 flight lines, 152 m line spacing, 91 m above ground level (AGL), 120 knots). Small Boy was approximately a 1 kt device detonated in July, 1962 as part of Operation Sunbeam that focused on testing small "tactical" nuclear warheads. It should be noted that the Small Boy yield was much less than Sedan's (104 kt) which suggests the Small Boy detection probabilities for Cs-137 should be lower than Sedan's and that these data can ultimately be used to quantify the contamination levels. Figure 10.3 shows two methods commonly used to process gamma-spectroscopy data. The first method is based on the Cs-137 ROI and the second is based on a more sophisticated analysis known as Noise Adjusted Singular Value Decomposition (NASVD), [2]. The actual Cs-137 deposition pattern was confirmed by ground-based measurements conducted by the Department of Energy (DOE) and is shown by the isopleth.

The image based on the Cs-137 ROI clearly shows an area in the northeast corner that suggests Cs-137 contamination and it is not present in the NASVD image. The NASVD method was able to eliminate the false-positive observed in the ROI

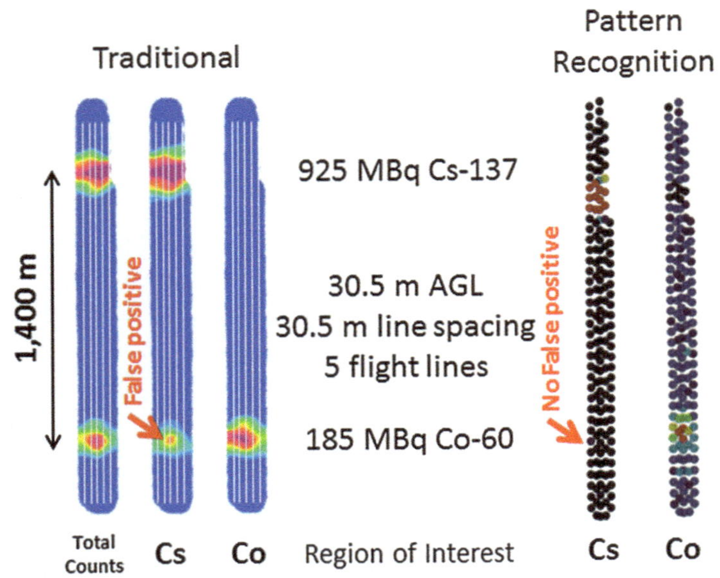

Fig. 10.2 Specificity of the PR method between Cs-137 and Co-60 traditional gamma-spectrometry methods (*contour*; *left* side) and pattern recognition results (*colored dots*; *right* side). *Color codes* for Cs-137 are the same as in Fig. 10.1. The *color codes* used for the Co-60 results are (1) *red* – strong active classification; (2) *yellow* – medium active; (3) *green* – weak active; (4) *light blue* – very weak active; (5) *dark blue* – weak inactive classification; (6) *purple* – medium inactive; and (7) *black* – strong inactive. Insufficient data are currently available to establish confidence levels for the Co-60 results

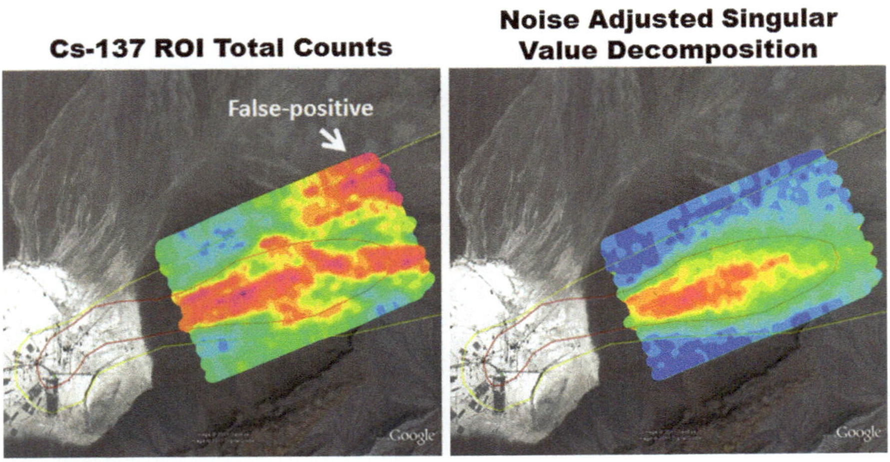

Fig. 10.3 ASPECT data contoured for Cs-137 showing fallout pattern from the Small Boy nuclear test conducted in 1962. The isopleth was verified by the DOE using ground-based data. Aerial data (20 flight lines, 91 m AGL; 152 m line spacing; 6 miles^2; 120 knots) were collected in 2010. The traditional ROI method (*left image*) is more prone to false-positives than the NASVD technique (*right image*)

method, because it analyzes the data for all independent spectral shapes and a Cs-137 shape was not present in this area. This is one key advantage of the NASVD method over the traditional ROI method. Two key differences between the NASVD and PR methods is the amount of computational time needed to implement each technique and the degree of user interaction required. The PR method produces near real-time results during the survey through an automated and objective process whereas the NASVD approach takes longer to produce a product and it is not easily automated due to the subjective nature of the analysis.

The Small Boy area was also surveyed by the DOE helicopter using a different set of flight parameters and volume of NaI crystals (11 flight lines, 305 m line spacing, 152 m AGL, 75 knots, and 25 L NaI). Figures 10.4 and 10.5 display results from the application of the Cs-137 PR model to the ASPECT and DOE data, respectively. Several key findings are apparent. First, there is a high degree of correlation between the PR results and the location of Cs-137 deposition as seen within the isopleth. The false-positives associated with the traditional ROI approach are also not apparent. Second, there are remarkable similarities between the data collected by the EPA ASPECT fixed-wing aircraft and the DOE helicopter. This is noteworthy because the Cs-137 PR model was derived solely from background and Cs-137 reference data collected by the ASPECT aircraft which was only equipped with 16.7 L of NaI at that time. The classification model successfully characterized the Small Boy fallout pattern using two different aircraft configurations (fixed-wing and helicopter) under different detector setups (16.7 L NaI vs 25 L NaI) at two altitudes (91 m AGL vs 152 m AGL) and different speeds (120 knots vs. 75 knots). This suggests that

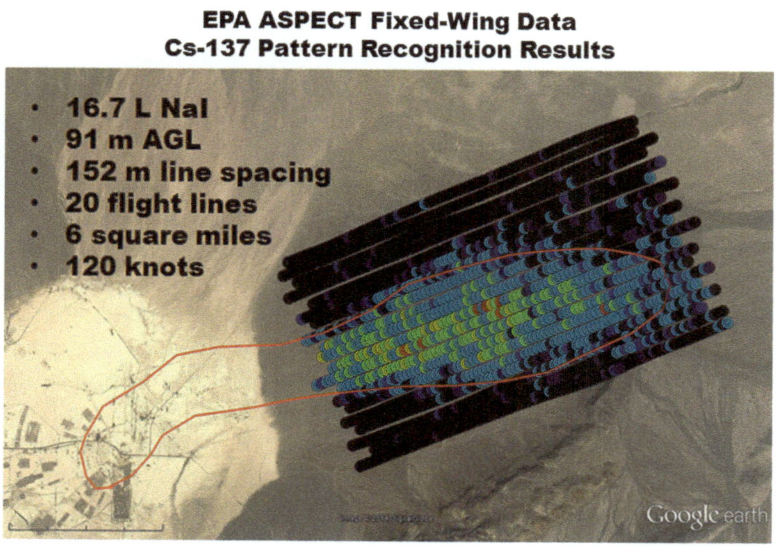

Fig. 10.4 Cs-137 pattern recognition results applied to the survey data collected by the EPA ASPECT fixed-wing aircraft in 2010. The *color codes* used are the same as those displayed in Fig. 10.1

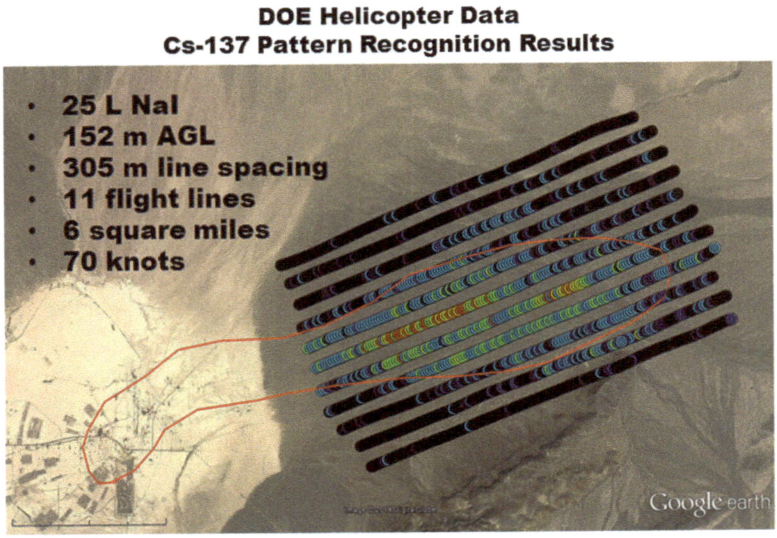

Fig. 10.5 Cs-137 pattern recognition results applied to the survey data collected by the DOE helicopter in 2010. The *color codes* used are the same as those displayed in Fig. 10.1

the PR method is globally applicable and not dataset dependent. It can be applied to any dataset collected by virtually any type of aircraft without the need for aircraft-specific calibration parameters, provided the detector types are consistent and the signal-to-noise criteria are met.

These results demonstrate that the PR method can be used to process gamma-spectral data to produce radionuclide-specific maps with comparable or better sensitivity at near real-time speeds in an objective process. These are critical components to the EPA ASPECT Program whose primary mission is emergency response and environmental characterization. It does not negate the value of the traditional methods since they provide adequate information depending on the purpose of the survey where time is not a critical factor.

This method is also useful for large-scale incidents like Fukushima, where a clean background dataset was difficult to obtain near the targeted survey areas. Traditional methods rely on a background that is typically collected near the survey site and usually consists of hundreds of data points. If these data have to be collected farther away from the contaminated sites, the acquired backgrounds may not be a good representation and can introduce additional uncertainty in the results. The PR coefficients were derived from tens of thousands of background data points from multiple locations and at various altitudes. Therefore, the automated processing does not require the need to collect background data during a response to properly characterize the spread of contamination.

10.5 Future Research

Research to refine and improve the radionuclide-specific PR coefficients continues with special emphasis on:

1. creating additional radionuclide-specific PR coefficients and mixes of radionuclides that pose risks to national security,
2. improving synthetic data generation to better account for Compton scattering while reducing the need to collect actual data over sources,
3. increasing the background spectral database to include more altitudes at more locations throughout the United States. This reduces the potential for false-positive detections and may improve overall detection sensitivities,
4. studying the application of using these PR coefficients on data collected from any aircraft platform that has the same detector configuration (e.g., crystal types), and
5. quantifying the ground concentrations based on discriminant scores.

10.6 Conclusions

The primary goal of the EPA ASPECT Program is to provide actionable intelligence about potential chemical and radiological threats to decision makers within minutes of data collection via the aircraft satellite communication system. The program has successfully used the PR method to characterize immediate threats from hazardous chemicals in the environment for many years and is beginning to implement this capability for gamma-emitting radionuclides. Classification models for Cs-137 and Co-60 have been developed and successfully tested using point-source and distributed-source configurations.

The PR method is an important addition to the suite of algorithms used by the ASPECT Program because of its ability to (1) eliminate traditional aircraft calibration protocols (i.e., stripping coefficients, radon corrections, etc.) while still being able to provide radionuclide-specific detection and mapping capabilities, (2) eliminate false-positive detections, (3) produce results in near real-time, and (4) allow independent flight operations that do not require specific flight line spacing or altitudes. This resistance to the effect of altitude arises from the background suppression methodology incorporated into the PR method. Because of loss of emission signal intensity through Compton scattering, however, signal-to-noise requirements limit the range of altitudes to approximately 350 m.

With proper training, any aircraft (e.g., helicopter or fixed-wing) outfitted with similar detection technology from which the PR models were developed (e.g., NaI detectors) can obtain this radionuclide-specific detection capability without the calibration requirements or technical expertise normally reserved to special programs. Thus, this method has the potential to significantly improve the nation's capacity to detect and characterize gamma-emitting radionuclides by enabling direct implementation by state and local authorities.

Acknowledgement The authors of paper would like to acknowledge our Chief Pilot, Beorn Leger (Airborne ASPECT Inc.), for flying the EPA surveys with great precision and accuracy.

References

1. ASPECT (2014) Retrieved August 21, 2014, from http://www2.epa.gov/emergency-response/aspect
2. Hovgaard J, Grasty RL (1997) Reducing statistical noise in gamma-ray data through principal component analysis. In: Gubins AG (ed) Proceedings of exploration 97: fourth decennial conference on mineral exploration, Toronto, pp 753–764
3. International Atomic Energy Agency (1991) Airborne gamma ray spectrometry surveying, Technical report series no. 323. International Atomic Energy Agency, Vienna
4. International Atomic Energy Agency (2003) Guidelines for radioelement mapping using gamma ray spectrometry data, IAEA-TECDOC-1363. International Atomic Energy Agency, Vienna
5. Kaltenbach TF, Small GW (1991) Development and optimization of piecewise linear discriminants for the automated detection of chemical species. Anal Chem 63:936–944
6. Small GW, Kroutil RT, Ditillo JT, Loerop WR (1988) Detection of atmospheric pollutants by direct analysis of passive Fourier transform infrared interferograms. Anal Chem 60:264–269
7. Wan B, Small GW (2008) Airborne passive Fourier transform infrared remote sensing of methanol vapor from industrial emissions. Analyst 133:1776–1784

Chapter 11
RadNet National Air Monitoring Program

Ronald Fraass

Abstract The United States Environmental Protection Agency has operated a variety of national radiological air monitoring systems for more than four decades. The current system, RadNet, operates approximately 140 fixed monitoring systems across the United States including Alaska and Hawaii. The system provides near-real-time gamma spectroscopy data and limited gross beta radiation data as detected on a 4 in. air filter on a high volume air sampler. Filters are analyzed hourly and results are sent electronically to an EPA laboratory for review. Any anomalous results outside of normal background variation are reviewed by an EPA scientist. In addition to the fixed monitoring systems, there are 40 deployable units that include low and high volume air samplers and electronically transmit ambient gamma exposure readings. The system provided continuous data to the public and interested scientists during the Fukushima release. RadNet is able to remotely detect typical gamma emitting isotopes at levels several orders of magnitude below protective action guidelines. In order to provide even more sensitive analytical results, RadNet air filters are sent to an EPA laboratory for analyses approximately twice weekly. Potential updates to the system include improved radiation detectors and improved communication devices for deployable RadNet systems.

11.1 Introduction

In the last decade, the United States Environmental Protection Agency saw a need to develop and institute a significant upgrade to the national radiation monitoring systems that it has had in place since the 1970s. The original monitoring network was primarily designed to detect fallout from early nuclear weapon testing by nation states. The new plan was defined in the 2005: *Expansion and Upgrade of the RadNet Air Monitoring Network, Volume 1 and 2, Concept and Plan* [1]. Prior to that plan, the EPA had 59 high-volume air monitoring systems located across the United

R. Fraass (✉)
US Environmental Protection Agency, National Center for Radiation Field Operations,
Las Vegas, NV, USA
e-mail: fraass.ron@epa.gov

S. Apikyan, D. Diamond (eds.), *Nuclear Terrorism and National Preparedness*,
NATO Science for Peace and Security Series B: Physics and Biophysics,
DOI 10.1007/978-94-017-9891-4_11

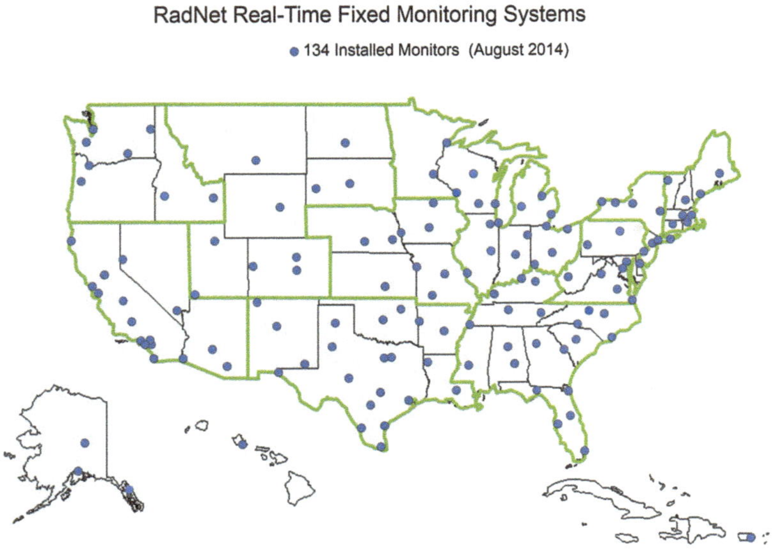

Fig. 11.1 Installed RadNet fixed location air monitors [4]

States. Those air monitoring systems were part of the Environmental Radiation Ambient Monitoring System (ERAMS) [2]. This system was different from the Comprehensive Test Ban Treaty monitoring network. Air filters from those sites were sent approximately twice weekly to the National Air and Radiation Laboratory (now the National Analytical Radiation Environmental Laboratory) for analysis. Data from the analyses were made available to scientists and the public via ERAMS data bases. That data is still being updated from the new RadNet system and includes data from the Chernobyl release. In addition to the air monitoring systems, RadNet also maintained a few sampling locations for rain water, drinking water, and milk. The updated air monitoring systems include hourly electronic data from gamma and beta detectors, a significant increase in the number of monitors (134 now installed), and 40 deployable systems with different characteristics from the fixed installation monitors. See Fig. 11.1 for locations of fixed monitors.

11.2 RadNet Mission and Objectives

RadNet's primary purpose is to track national or regional ambient levels of radioactive material in the air. This allows users to identify the degree and extent of contamination in the event of an emergency. The system operates continuously so that normal background levels and natural fluctuations are known. The system supports EPA's role in incident assessment and focuses on monitoring potential impacts to public health.

Radnet provides data quickly in the event of a radiation incident to decision makers, dispersion modelers, nuclear and radiation health experts, and the general public. Although the system might be the first to detect an incident, it is normally expected to provide data following a known incident such as Chernobyl or Fukushima. RadNet data helps determine large scale national impacts of an incident, timely data for modelers, and estimates of exposure rates to assist in protective action recommendations and dose reconstruction.

RadNet is not intended to be regulatory in nature. As such it is not used to monitor nuclear reactors or provide early warning of nuclear accidents. As a national level system, it is not intended for use in the immediate locality of an incident. Other federal, state, local, or tribal assets will be used to monitor at a specific incident site.

11.3 RadNet Basics

While this paper primarily address the air monitoring aspects of RadNet, the program has been and continues to be multi-media including water and milk. However; the milk program has recently been discontinued and will be assumed by other federal agencies. RadNet air monitoring systems are typically operated by volunteers from EPA, states, and locals. All air filter and iodine cartridge samples are analyzed at the National Analytical Radiation Environmental Laboratory (NAREL) in Montgomery, Alabama. The hourly and analytical data are available to the public.

RadNet air samplers (at fixed locations) operate at 60 cubic meters per hour and collect air samples on a polyester filter. A 2 in. by 2 in. sodium iodide (NaI) detector and a 600 square millimeter ion-implanted silicon detector make and transmit hourly measurements of the filter. Data from the sodium iodide detector is broken down into ten energy regions of interest for ease of data transmission. Staff at NAREL review the data if a computer algorithm indicated values are above typical background variation. There are two levels of alert depending on the size of the variance. If the count rates exceed the higher alert level, immediate notifications are sent to staff. While normal data flow only includes counts in the ten regions of interest and the beta count, staff are able to download the full NaI spectrum for analysis. An example of the spectrum from a calibration run is shown in Fig. 11.2. All air filters received by NAREL are screened for gross beta and undergo gamma analysis if above typical background levels.

The RadNet air monitoring systems at fixed locations provide excellent sensitivity for radionuclides of interest. Typically more than 90 % of the systems are operational at any given time. Remote sensitivity for several isotopes of interest is approximately 2 nCi (74 Bq) on the filter as shown in Fig. 11.3. For those filters analyzed at the laboratory, the sensitivity is approximately two orders of magnitude better. During the response to Fukushima, a few filters were counted for 5,000 min to ensure that all necessary isotopes were included in the gamma spectroscopy isotope analytical library. Yearly, over one million electronic data sets are reviewed and over

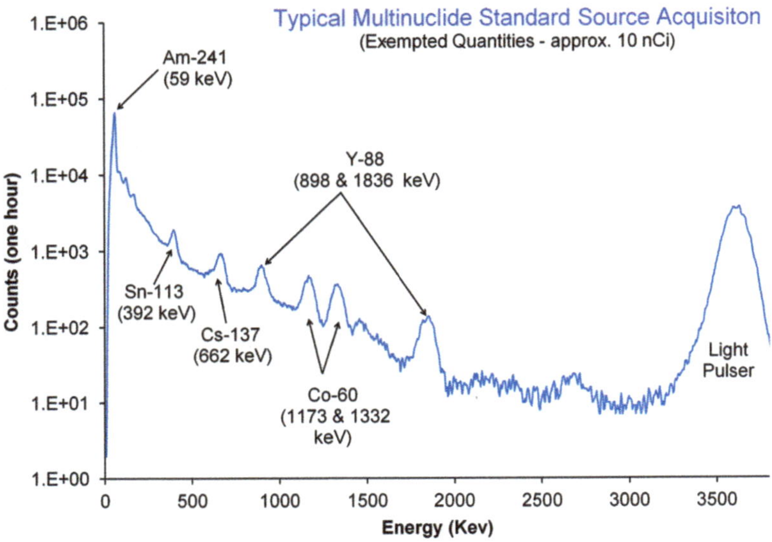

Fig. 11.2 Remote calibration spectrum from NaI Detector

Isotope	1-hour MDA*	PAG DRL †
Am-241	13.2 nCi	114 nCi
Cs-137	1.2 nCi	1,440,000 nCi
Co-60	2.0 nCi	222,000 nCi
Cs-134	1.1 nCi	960,000 nCi
Ir-192	1.6 nCi	1,620,000 nCi
Sr-90	50 nCi	38,400 nCi

*One hour measurement Minimum Detectable Activity at 95% confidence level
†DRL equivalent to Protective Action Guide

Fig. 11.3 Remotely measured sensitivity of NaI Detector compared to derived response level

14,000 laboratory analyses are performed. An additional, highly sensitive, analysis for plutonium in the air is performed on a composite sample of ash from all filters collected from a specific monitor for that year.

Data from the early and current RadNet systems are available at:

http://www.epa.gov/radnet/radnet-data/
http://iaspub.epa.gov/enviro/erams_query_v2.simple_query

The fixed location monitors send data via four redundant means to the monitoring center including phone lines, internet, cellular modem, and satellite. The deployable monitors transmit their data via analog modem, satellite, or directly to a personal digital assistant attached via cable to the system.

Deployable RadNet monitors include a low-volume air monitor, a high-volume air monitor, and a pair of compensated gamma radiation monitors. The gamma radiation monitors are Genitron Gamma Tracers and can measure exposure rates from 10 nSv/h to 10 mSv/h. External exposure rates from the gamma detectors are typically sent every 15 min to the monitoring center at NAREL when the deployables are dispatched and operational. The low-volume air sampler normally uses both an air filter and a cartridge (activated charcoal or silver zeolite) for collecting particulates and iodine vapor. The high-volume air sampler only uses an air filter.

EPA maintains 40 deployable monitors ready for use in an incident. Currently 20 are stored at NAREL and 20 are stored at the National Center for Radiation Field Operations (NCRFO) at Las Vegas, NV. In the future, the majority will be stored and maintained ready for use at NCRFO.

11.4 RadNet Isotope Detections

Figure 11.4 shows the detectors and interior of a fixed RadNet monitor. There is no special shielding provided for the NaI detector. Because of this, it is able to detect radionuclides in its environment in addition to monitoring the air filter. Because of this, RadNet monitors have seen a number of isotopes in their vicinity. The first was the detection of a radiographer using a cobalt-60 source in a major city. The source was about a block away, but clearly identifiable. Because the count rate went back to normal after the source was secured by the radiographer, the situation was clearly not due to collection of radioactive material on the filter.

Since that first detection, RadNet has seen iodine-131 and technetium-99m from medical patients, cesium-137, iridium-192, and most recently a positron emitter (most likely fluorine-18.) Each of these instances caused a low level alert so that an EPA scientist evaluated the spectrum to identify the source and confirm that it did not indicate an airborne release.

11.5 Fukushima Response

Within hours of learning about the damage to the Fukushima reactors, RadNet systems were continuously monitored for any indication of the plume reaching the United States. Because the airborne concentration was not expected to be high

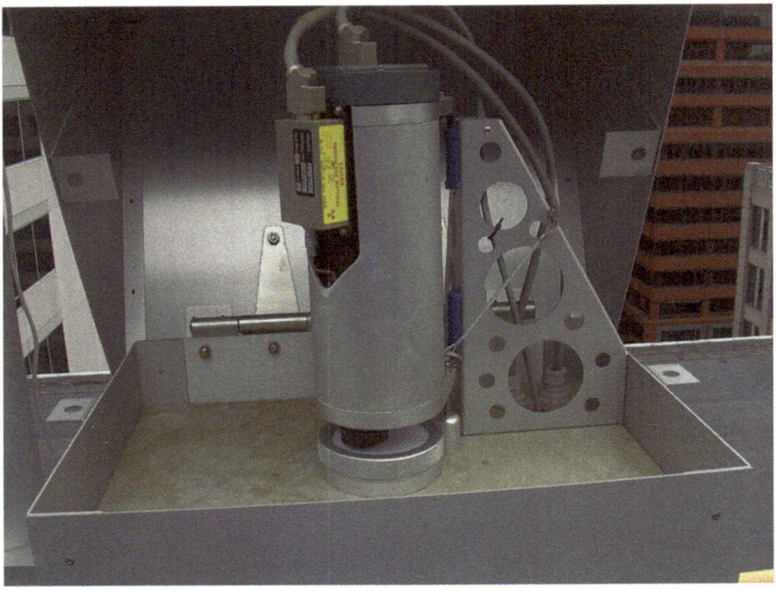

Fig. 11.4 RadNet fixed location monitor detectors

enough to be detected by the near-real-time NaI or beta detectors on the fixed location RadNet systems, NAREL scientists analyzed several of the first filters that might contain isotopes from the event. Normal gamma analyses are for 3,000 min but a few of those early filters were analyzed for 5,000 min. Isotopes detected included Cs-134, Cs-137, I-131, I-132, and Te-132. As the plume moved across the United States, it was also detected in rain water samples. NAREL analyzed each rain water sample rather than compositing them monthly while the plume was present. During the entire time that RadNet systems were monitoring the event, there was only about 3 h where isotopes presumed to be from Fukushima were detected remotely on an air sample. The filter was analyzed remotely and approximately 1 nCi (37 Bq) of radioactive iodine was detected. After filter change, no further radioactive iodine was detected. The remote analysis was later confirmed by laboratory analysis at NAREL.

Discussions with other federal, state, and industry representatives indicated that they were also able to detect the plume using laboratory analysis of air samples. This raised the question of how to properly inform the public. RadNet fixed location monitors were not seeing the plume above background yet more sensitive analyses showed its presence at levels extremely far below any protective action level [3]. The added radioactive material was completely lost in the normal background variation seen by the RadNet monitors.

11.6 Upgrades Being Considered

For the fixed location RadNet systems, EPA is considering potential upgrades to the NaI detector to improve resolution, to the beta detector to reduce radiofrequency interference, and increased data transmission rates either for satellite or modem systems.

For the deployable RadNet systems, EPA is considering a significant change to the system that collects the gamma exposure, air sampler flow rates, and other data. This work is being done in conjunction with the US Department of Energy. Another potential change may be the elimination of the low and high-volume air samplers from the system as there are numerous such samplers in the EPA radiation response inventory. These proposed changes would reduce the weight and physical footprint of the deployable monitors.

11.7 Conclusion

The 134 fixed location and 40 deployable RadNet air monitoring systems have significantly improved the ability of the Environmental Protection Agency to monitor actual and potential large area releases of radioactive material in the air. The original system of 59 air monitors provided significant data to the agency following the Chernobyl incident. The new system provided hourly data to the scientific community and public during the Fukushima incident. While the airborne concentrations in the United States were far below any levels requiring protective actions, the public wanted to see for themselves. Data hits on the RadNet website prior to the incident averaged perhaps 30 a day. During the Fukushima event, that rose to several million hits a day. Clearly the public wants almost immediate access to information that they consider important to their health. RadNet was able to provide that.

References

1. U.S. Environmental Protection Agency, Office of Radiation and Indoor Air (2005) Expansion and upgrade of the RadNet air monitoring network; volume 1&2; Concept and plan, U.S. Environmental Protection Agency, Washington, DC
2. U.S. Environmental Protection Agency, Office of Radiation and Indoor Air, National Air and Radiation Environmental Laboratory (1988) Environmental Radiation Ambient Monitoring System (ERAMS) Manual (EPA 520/5-84-007, 008,009). Montgomery, U.S. Environmental Protection Agency, Washington, DC
3. U.S. Environmental Protection Agency, Office of Radiation Programs (1992) Manual of protective action guides and protective actions for nuclear incidents (EPA-400/R-92-001). U.S. Environmental Protection Agency, Washington, DC
4. Askren D (2014) Image of 136 fixed RadNet monitor locations (unpublished)

Chapter 12
Radiological Emergency Response and Preparedness at the Remote Sensing Laboratory

Sanjoy Mukhopadhyay and Richard Maurer

Abstract This article describes the capabilities and types of radiological emergency responses and preparedness provided by the Remote Sensing Laboratory (RSL). RSL is owned and operated by National Security Technologies, LLC. RSL provides the National Nuclear Security Administration with a broad range of scientific, technological, and operational disciplines with core competencies in emergency response operations and support; remote sensing; and applied science and technologies in support of radiological emergency preparedness, and radiological incident response.

RSL operates out of two bi-coastal locations, one at the Nellis Air Force Base, Las Vegas, Nevada and the second at the Joint Base Andrews, Andrews Air Force Base, and Maryland. In 1976, The U.S. Department of Energy established an Aerial Measurements Operations at Andrews Air Force Base in Maryland to provide scientific and technical support to counterterrorism efforts during U.S. bicentennial events in Washington, D.C. With a location on each coast, RSL has served for over 50 years as a valuable national asset for nuclear emergency response and remote sensing capabilities.

RSL remote sensing capabilities include:

I. Radiation detection, monitoring, surveillance, and analysis
II. High-speed data telemetry
III. Secure mobile communications
IV. Geographic information systems
V. Photography and videography
VI. Multi-spectral and hyper-spectral imagery

S. Mukhopadhyay (✉) • R. Maurer
National Security Technologies, Joint Base Andrews, MD, USA
e-mail: MUKHOPS@nv.doe.gov

S. Apikyan, D. Diamond (eds.), *Nuclear Terrorism and National Preparedness*, NATO Science for Peace and Security Series B: Physics and Biophysics, DOI 10.1007/978-94-017-9891-4_12

12.1 Background

The Remote Sensing Laboratory (RSL) – Andrews Operations was established in 1976 before U.S. Bicentennial events in Washington, D.C. with the important mission to provide aerial radiological measurement services and radiological emergency response and consequence management assistance to the greater Washington area. With heavy emphasis on Aerial Measuring System (AMS) the institute was called Washington Aerial Measuring Operations (WAMO). WAMO used to provide routine detailed aerial surveys of nuclear power plants (NPP) for the Nuclear Regulatory Commission (NRC). With the creation of National Nuclear security Administration (NNSA) in 2000 RSL was integrated with the US Department of Energy's Office of Emergency Management (OEM) as part of National Center for Nuclear Security (NCNS)'s management and operations contract owned and carried out by National Security Technologies, LLC at the Nevada National security Site (NNSS). Currently The Remote Sensing Laboratory (RSL) creates advanced technologies for emergency response operations, remote sensing, counterterrorism, and radiological incident response. RSL emergency response teams are on-call 24 h a day and are specially trained and fully equipped to respond to a variety of radiological situations. We maintain the capability to deploy domestically and internationally in response to nuclear threats involving the loss, theft, or release of radioactive material that might occur in nuclear power plant accidents, nuclear terrorist incidents, NASA launches, and transportation accidents. The NNSS helps ensure the security of the United States and its allies by supporting the stewardship of the nuclear deterrent, providing emergency response capability and training, and contributing to key nonproliferation and arms control initiatives. We execute unique national-level experiments, support national security customers through work for others, manage the legacy of the Cold War nuclear deterrent, and provide long-term environmental stewardship for site missions.

The RSL has served for over 50 years as a valuable national asset for nuclear emergency response and remote sensing capabilities.

12.1.1 Emergency Response Capabilities

- National Aerial Measuring System (AMS) Reach back Center
- Search Management Center
- Federal Radiological Monitoring and Assessment Center

RSL emergency response teams are on-call 24 h a day and are specially trained and fully equipped to respond to a variety of radiological situations. They maintain the capability to deploy domestically and internationally in response to nuclear threats involving the loss, theft, or release of radioactive material that might occur in nuclear power plant accidents, nuclear terrorist incidents, NASA launches, and transportation accidents.

As part of its emergency response, the RSL provides:

- Crisis response and consequence management assistance

The crisis response capability includes the National Radiological Advisory Team (NRAT) and Radiological Assistance Program (RAP) that deploys personnel and detection equipment to support vehicle-based and handheld operations. The team utilizes high-resolution gamma and neutron detectors and is able to identify specific radioisotope. The Search Management Center (SMC) is a new enterprise which has recently been added to the RSL's emergency response capabilities. The SMC provides federal, state, local tribal and other first-responder communities with tools and direction to assist in wide-scale radiological search operations. It also enhances situational awareness for complex missions. The SMC core capabilities include:

A search planning tool for calculating time and coverage based on available resources
An instrument/personnel tracking system which includes smart-phone devices providing alarm, spectral data, and photographic telemetry as well as verbal and non-verbal field communications
A secure, web-based display system for viewing by mission commanders and other decision makers
A status module for monitoring critical search operations information

- Emergency communications networks
- Logistics and operations support
- Training and exercise planning and execution

12.2 Consequence Management (CM) and Federal Radiological Monitoring and Assessment Center (FRMAC)

CM's mission is to assist the state, tribal and other local governments with the evaluation of a radiological incident/accident. The RSL maintains a trained team of radiological field experts ready to respond on short notice to provide monitoring sampling analysis and assessment of field data. The RSL CM response teams also form the core of FRMAC in the event that a FRMAC is activated in response to a radiological incident. The recent development of the electronic Federal Radiological monitoring and assessment Center (eFRMAC) allows emergency response assets to gather more field data faster with greater reliability to better support decision makers. The capability handles all aspects of data acquisition, management, analysis and dissemination via secure protocols and standards of data communication. The key features and benefits of eFRMAC include:

- Flexible data collection clients
- Multi-path communication devices

- Robust transactional synchronization between home team and field assets
- Dynamic data management
- Connections to existing data repositories
- Expedited data entry for first responders

12.3 Emergency Communication Network (ECN)

NNSA's Emergency Communications Network (ECN) provides decision makers at NNSA with the capability to exchange real-time voice, data, and video information for managing emergency situations that involve NNSA assets and interests.

The Emergency Communications Network is a multi-faceted communications network providing classified and unclassified voice, video, and data communications between NNSA headquarters and approximately 55 remote sites and mobile units via dedicated leased lines and satellite transmission.

The Mobile Emergency Communications Network provides a portable dynamic communications capability for NNSA emergency response assets, with full connectivity to the ECN and possibly other networks. The Mobile ECN also provides satellite backup capability for the ECN and Home Team terrestrial circuits.

The Mobile ECN is capable of classified and unclassified data, voice, and video; wide-band satellite connectivity between NNSA emergency response assets and Home Teams; and simultaneous support of multiple deployed systems.

The ECN provides support to the Department of Energy's Emergency management, Emergency Response and International Emergency Management Cooperation missions with engineering design and implementation of world class communication networks and Emergency Operations Centers

- National and international
- Classified and Unclassified
- Wide area network (WAN) and local area network (LAN)
- Data, voice, video and video conferencing
- Dynamic scalable to meet a variety of emergency response mission requirements

Engineering design and implementation of mobile and satellite gateway emergency communications systems for:

- National and international emergency communications
- Continuity of operations (COOP)
- The strategic petroleum reserve (SPR)
- Expandable to support large scale or dispersed events via microwave and fiber optics lines
- Data, voice, video and video conferencing systems
- Satellite communications

Geographical information system (GIS) location-based situational awareness and decision support products and resources are listed below:

- GIS data acquisition and analysis
- GIS emergency technology
- GIS database management
- Customized map products

12.4 Aerial Measuring System (AMS)

AMS provides a rapid and comprehensive worldwide aerial measurement, analysis, and interpretation capability in response to nuclear/radiological emergencies resulting from threats, accidents, deliberate attacks, or lost/stolen materials. Detection sensitivities and response times are sufficient to assess health and welfare impacts on population to the levels outlined in the nationally and internationally adopted protective action guides. The system is dependable, standardized, authenticated, and is based on state-of-the-art technologies in aerial detection, data processing, telemetry, and systems integration.

The Remote Sensing Laboratory located at the Andrews Air Force Base maintains a Bell-412 helicopter and a B-200 fixed-wing aircraft, equipped with radiological detection systems and qualified pilots ready to deploy in the NCR with a short notice. The helicopter flies at a nominal speed of 70 knots at an altitude of 150 ft. above ground level (AGL) and can perform very detailed "grid-search" or "combing" operations, searching, locating discrete radiological sources, and/or providing characterization of large contamination areas. The helicopter can fly about 2.5 h before refueling. The fixed-wing flies at a nominal altitude of 1,000 ft. with a speed of 240 knots, covering a larger area in a shorter time. It is predominantly used for plume tracking, locating large discrete sources, and providing a "quick look" contamination footprint in large-scale nuclear or radiological accidents/incidents. Figure 12.1 shows how the helicopter flies parallel grid-lines to characterize and produce a radiological map of a large ground deposition.

12.5 International Emergency Management Cooperation (IEMC)

Cooperation and Emergency Management Planning with International Partners – The RSL provides scientific, engineering, technical advice, services, products, logistics and administrative assistance to the NNSA IEMC program in support of international outreach and global initiatives.

- Training and collaboration between countries on emergency management support
- Enables early detection of suspect materials before they are shipped to the United States
- Cooperation and sharing of equipment and information

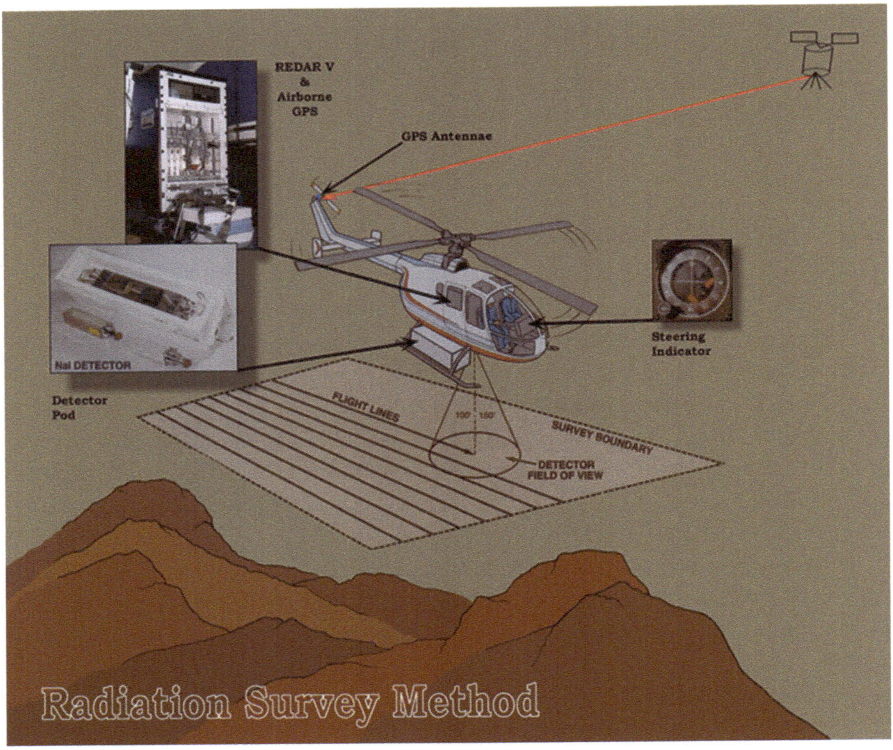

Fig. 12.1 A pictorial description of how to fly parallel grid lines for ground deposition mapping

12.5.1 Japan Earthquake/Tsunami Response

- Provided AMS and CM Response Team support to the U. S. Military, the U.S. Department of State and the Government of Japan following the Fukushima Dai-ichi Nuclear Power Plant accident
- Field operations center was located at Yokota Air Base
- Collected assessed and reported daily radiation measurements taken from an air- and ground-based platforms
- CM Home Team provided 24/7 support to the field teams and the Nuclear Incident Team (at DOE HQ)
- Maintained a 10-week-ling deployment with occasional staff rotations and team augmentation.

The Department of Energy/National Nuclear Security Administration's (DOE/NNSA) Aerial Measuring System (AMS) deployed personnel and equipment to partner with the U.S. Forces in Japan (USFJ) to conduct multiple aerial radiological surveys. These were the first and most comprehensive sources of actionable information for U.S. interests in Japan and provided early confirmation

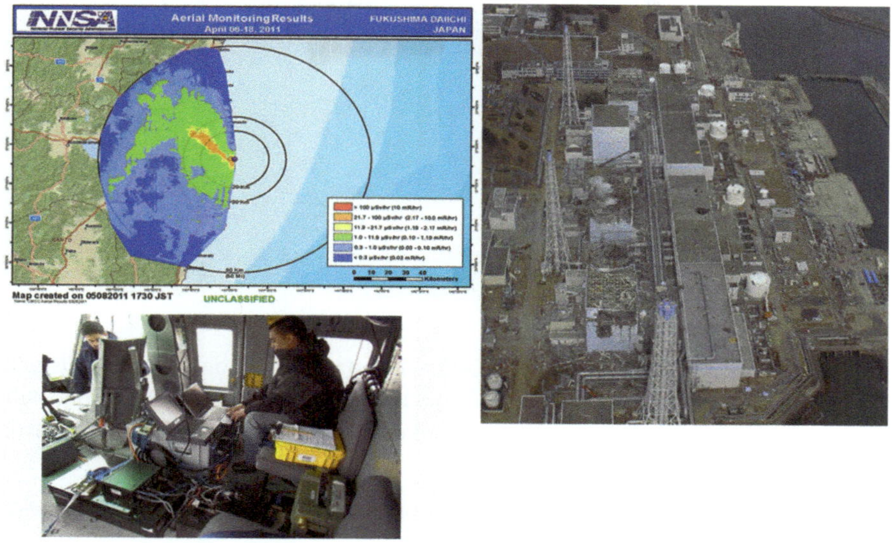

Fig. 12.2 Clockwise the final radiological map showing the extent of ground deposition around the Power plant. The aerial photo of Unit 2 at the Dai-ichi installation (*right*). Equipment setup and personnel in a helicopter in preparation before a flight

to the Government of Japan as to the extent of the release from the Fukushima Daiichi Nuclear Power Plant. Many challenges were overcome quickly during the first 48 h, including installation and operation of Aerial Measuring System equipment on multiple USFJ aircraft, flying over difficult terrain, and flying with USFJ pilots who were unfamiliar with the Aerial Measuring System flight patterns. These factors combined to make for a programmatically unanticipated situation. In addition to the challenges of multiple and ongoing releases, integration with the Japanese government to provide valid aerial radiological survey products that both military and civilian customers could use to make informed decisions was extremely complicated. The Aerial Measuring System Fukushima response provided insight into addressing these challenges and gave way to an opportunity for the expansion of the Aerial Measuring System's mission beyond the borders of the U.S. (Fig. 12.2).

12.6 Applied Science and Engineering

The radiological Emergency response mission is supported by a cadre of scientists, engineers and technicians at the RSL. Each is a member of deployable team and as such has gained significant experience in the development, application and execution of mission specific equipment. The equipment developed and produced at the RSL aids several NSA missions that include nuclear counterterrorism and radiological disaster consequence management

- Radiation detector system design
- Data acquisition and analysis software development
- Radiation detector networking
- Electronic board design
- Fabrication
- Telemetry integration

12.7 Homeland Security and Counterterrorism Solutions

The RSL supports nation's counterterrorism efforts with customized products and prototyping of equipment. Focused on rapid turn-around and advanced technology solutions, the RSL specializes in unique technological disciplines in counterterrorism including

- Special instrumentations for active and passive electromagnetic detection elements
- Port of entry radiation detection systems
- Sensor development, testing and application verification
- Real-time mission support
- Instruction and proficiency training on specialized and mission applications

12.8 Innovation – Radiological Mapper

In radiological emergency response operations, it is essential that the observed radiation data are associated with the geospatial position of the search area and overlaid on a map. Several of the currently available COTS and GOTS search equipment have the capability of routinely collecting and displaying data using GPS coordinates on a geo-referenced map. However, an increasing number of situations call for radiological data to be collected in areas where GPS receivers have limited or no signal strength, such as within the confines of a building. Hardened bunkers, inner decks of a cargo ship, and many other isolated structures are examples of GPS-denied locations. In such areas it is not easy to associate measurements with position. The usual procedures for accurately measuring position without the use of GPS (i.e., surveying) are painstaking and slow. Consequently there is a strong desire to have:

1. A metrology tool that can continuously record position even in the absence of GPS. Such a tool could then be coupled with radiological measuring equipment and provide the capability to conduct indoor surveys with radiological equipment.
2. A software tool that can project radiological data on a picture or map of the surveyed area by following the spatial correlation of the radiological data.

This white paper purports to develop the software tools required to perform the mapping of radiation data in a GPS-denied area, make the system more selective by introducing gamma-ray spectral features, and incorporate a robust metrology tool that can continuously record position.

One of the goals of the project will be to determine the technologies and equipment presently available that provide a pedestrian the ability to accurately record position in an automated fashion, suitable for integrating with other data collection systems. This work will be divided into two distinct parts. The first part will survey the available technologies and evaluate which technologies are suitable for our needs; the second part will experimentally evaluate a set of such devices to find the most appropriate one that best suits the search applications.

The latter part of the project will integrate the neutron and gamma-ray data from the backpack-type search system with spatial location data. The radiation data will then be projected onto a background map, which then would allow for the detection of sub-threshold sources, document search path, and present the data for follow-on assessment. The project execution involves the designing of the integration software into a platform that is more suitable for field operations. An important specification is that the package would be scalable to more advanced search applications, including advanced, integrated analysis. The platform operates by manual input of search location, which is then synchronized with the streaming backpack data. This system's features include the ability to edit the data for errors, enhanced safety features during operation, and improved usability. The future capability to be integrated with a data analysis routine is a driving force during the system development. Thus, MATLAB was chosen over other packages because of its ability for stand-alone execution, rapid-prototyping, and existing analysis capability.

Finally, by incorporating gamma-ray energy spectroscopy features of radioactive materials of interest, the search system can be made more selective. This would also reduce false positive and negative alarms. Several algorithms like Sequential Probability Ratio Test, maximum likelihood optimization, covariance spectroscopic analyses, characterizing, and parameterizing urban radiological background will be used to improve the real-time search data output.

Figure 12.3 displays an example of the real-time performance of the pedestrian Mapper.

12.9 Multi-path Communication Device (MPCD)

The MPCD is a versatile communications system that can relay data from the field to command posts via fixed web assets. Features of the system include:

- Transmission over several commercial or standalone communications mechanisms. The system pictured in Fig. 12.1 can transmit over two types of cellular data networks, satellite and RF. Transition between the different communication modes happens instantaneously if the current mode becomes unavailable.

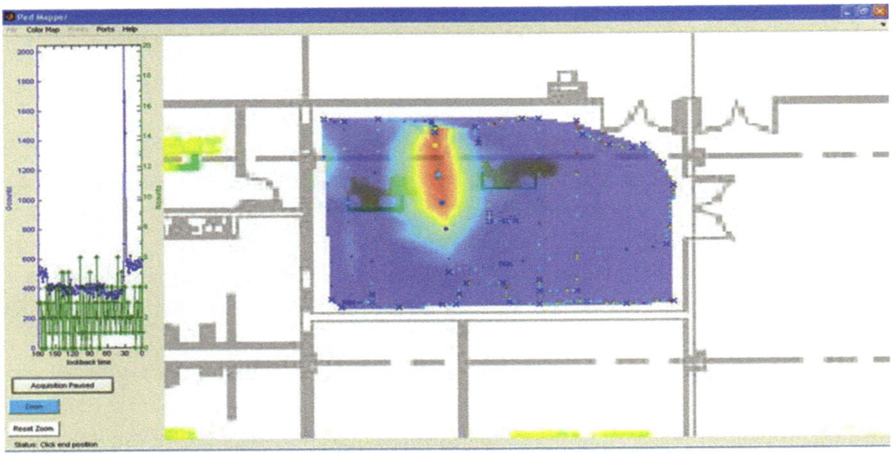

Fig. 12.3 Real-time Radiological Mapper. On the *left* is shown a strip chart consisting of both gamma-ray and neutron counts. On the *right* the *blue dots* are markers made along the search path, colored contours (*blue* – benign to *orange* – radioactively "hot") show real-time isopleths as the searcher moves around – they change with time

- Communications link that can be adjusted not only on most-capable status but also on least-cost considerations.
- Ability to access power from battery, AC or a "cigarette lighter" connection (12 VDC).
- An onboard GPS.
- Ability to autonomously receive, package, buffer, archive, and transmit data to a mobile base station or via the Internet to fixed web assets.
- Ability to be used with several types of systems from mobile vehicle, static sensors to aerial platforms.
- Easy integration into existing detection assets through serial, Ethernet, Bluetooth or 802.11 b/g so that they can transmit through MPCD.

The MPCD is a vehicle-mounted device that provides a communication pathway for field measurement teams to transmit collected radiological measurements and sample data for analysis by monitoring and assessment. This capability speeds up the process of data driven decision making during an incident. The user (operator) enters these radiological measurements and collected samples into a tablet PC running the Digital Field Monitoring (DFM) software which communicates wirelessly (802.11) to the MPCD.

The MPCD transmits these measurements via satellite, cellular, and mesh networking technologies. It does so automatically with no user intervention. This data is collected at the Deployed Command Center (DCC) (if deployed) or at the Remote Sensing Laboratory in real time and can be viewed immediately by any user with an account and an internet connection at the Radiological Assessment and Monitoring System (RAMS) website https://frmac.oem.doe.gov.

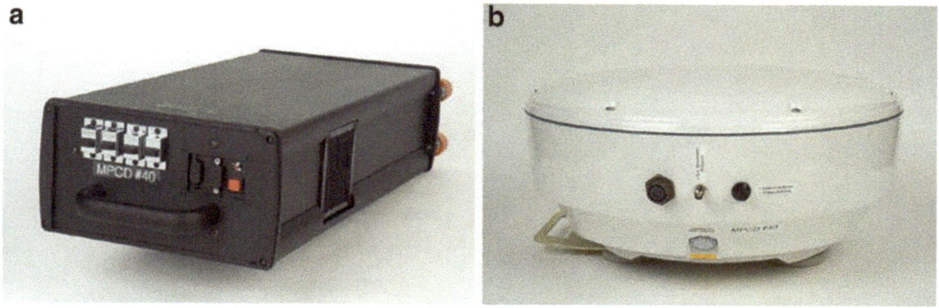

Fig. 12.4 (**a**) Indoor unit (IDU). (**b**) Outdoor unit (ODU)

12.9.1 Attributes

- Physical Characteristics
  - IDU (Indoor Unit) shown in Fig. 12.4a
  - Size: 19 × 9 × 5 in.
  - Weight: 20 pounds
  - ODU (Outdoor Unit) shown in Fig. 12.4b
  - Size: 18 × 18 × 11 in.
  - Weight: 13 pounds
- Power
  - Input 12 VDC smart controller
  - Battery Backup
  - Operating Time: Indefinite
  - (Vehicle controlled)
- Communications Pathways Outbound (Commercial)
  - Globalstar Satellite
  - AT&T Broadband
  - Verizon Broadband

Acknowledgements This manuscript has been authored by National Security Technologies, LLC, under Contract No. DE-AC52-06NA25946 with the U.S. Department of Energy and supported by the Site-Directed Research and Development Program. The United States Government retains and the publisher, by accepting the article for publication, acknowledges that the United States Government retains a non-exclusive, paid-up, irrevocable, world-wide license to publish or reproduce the published forms of the manuscript, or allow others to do so, for United States Government purposes.

Chapter 13
Evaluation, Improvement and Guidance for the Use of Local-Scale Emergency Prediction and Response Tools for Airborne Hazards in Built Environments. Cost Action ES100 – A European Experience

Tamir G. Reisin, Bernd Leitl, Silvia Trini Castelli, Kathrine Baumann-Stanzer, Patrick Armand, Fotios Barmpas, Spyros Andronopoulos, Klara Jurcakova, and all COST ES1006 Members

Abstract Releases of hazardous agents, such as in a RDD event, in complex built environments pose a tremendous challenge to emergency first responders and authorities in charge due to casualties potentially involved and the significant environmental impact. Air motions in built-up areas are very complex and adequate

T.G. Reisin (✉)
Israel Atomic Energy Commission, Tel Aviv, Israel
e-mail: tgreisin@gmail.com

B. Leitl
Meteorological Institute, University of Hamburg, Hamburg, Germany

S.T. Castelli
Institute of Atmospheric Sciences and Climate, National Research Council, Torino, Italy

K. Baumann-Stanzer
Central Institute for Meteorology and Geodynamics, Vienna, Austria

P. Armand
CEA, DAM, DIF, Arpajon, France

F. Barmpas
Laboratory of Heat Transfer & Environmental Engineering, Aristotle University of Thessaloniki, Thessaloniki, Greece

S. Andronopoulos
National Centre for Scientific Research "Demokritos", Institute of Nuclear and Radiological Sciences and Technology, Energy and Safety, Attiki, Greece

K. Jurcakova
Department of Fluid Dynamics, Institute of Thermomechanics, Academy of Sciences, Prague, Czech Republic

S. Apikyan, D. Diamond (eds.), *Nuclear Terrorism and National Preparedness*, NATO Science for Peace and Security Series B: Physics and Biophysics,
DOI 10.1007/978-94-017-9891-4_13

modelling tools have to be applied properly in order to predict the dispersion of hazardous materials with sufficient accuracy within a very short time. Different types of tools are applied; however, it is not always clear the advantages and limitations of individual models and approaches. Therefore, it is of an exceptional interest to compile a detailed inventory of the different models and methodologies currently in use, to characterize their performance and to establish strategies for their improvement. The Action is a first cross-community initiative to join, to coordinate and to harmonize European efforts for a substantial improvement in the implementation of local-scale emergency response tools.

13.1 Introduction

The atmospheric dispersion models (ADM) represent a crucial part of local-scale emergency response tools (ERT) for tracking and predicting airborne hazards from accidental or deliberate releases (see Fig. 13.1). A major challenge is their application in complex topography and geometry, as in industrial or urban environments. Various modelling approaches are applied, from simple parametric models and Gaussian methods to Lagrangian dispersion models and advanced CFD-based modelling suites. The different methodologies have advantages and disadvantages

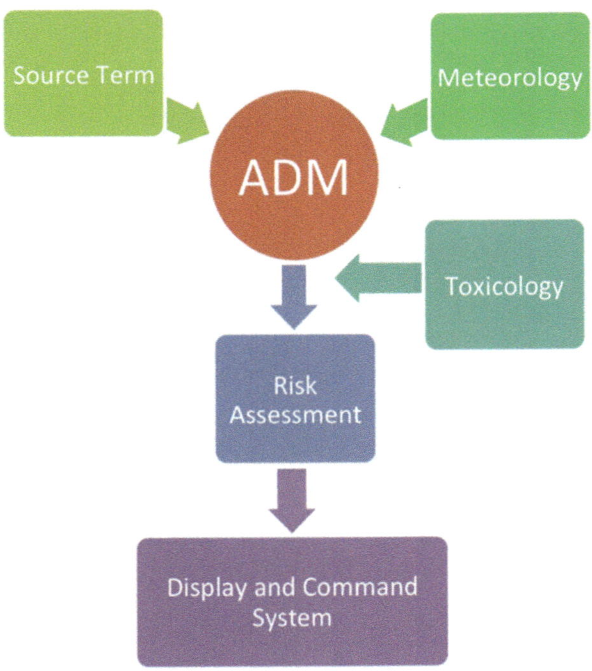

Fig. 13.1 The principal components of an ERT with the ADM at its core

with respect to their computational efficiency, accuracy, reliability of the results and many more. For any accidental release scenario, authorities can take different decisions and a variety of instructions can thus be provided to the emergency responders, depending on the simulation tools applied.

Modern societies face a large spectrum of potential threats connected to the release of hazardous agents into the atmosphere. These threats include the release of:

- Radioactive or nuclear materials (e.g. Fukushima, Japan, 2011),
- Toxic chemical products as gases or particles (e.g. Seveso, Italy, 1976),
- Pathogenic biological entities (e.g. Sarpsborg, Norway, 2005).

These CBRN threats might be realized as the result of accidents, or as a consequence of criminal or malevolent activities. Although these events are quite different in nature, effective emergency response to them requires similar atmospheric dispersion modelling and risk assessment tools.

The COST Action ES1006 was established with the main goals of evaluating the application of ADM in built environments, that is urban or industrial sites, and of assessing their integration in ERTs. The focus is on short-term and small-scale threats, which the emergency services are most often called to cope with. The Action activity is mainly aimed at:

- Elaborating a complete inventory of local threat scenarios and related modelling systems presently used, which could be of reference for local-scale airborne hazards modelling;
- Setting up a dedicated comprehensive and structured inventory of models suitable to local-scale accidental releases;
- Investigating the main gaps, deficiencies and limitations in presently available knowledge and models, identifying the directives for their improvement;
- Addressing the integration of airborne hazards modelling tools in ERTs for urban/industrial applications;
- Evaluating available models with an application-oriented approach, through validation against observations from qualified field and laboratory experiments and model inter-comparison.

The Action encompasses three major activities:

Working Group 1: Threats, Models and Data Requirements

- Compilation of test data sets available for systematic validation of local-scale dispersion models in complex urban environments
- Definition of an evaluation scheme for qualified reference data sets

Working Group 2: Testing, Evaluation and Further Development of Models

- Validation test runs (blind/no blind)
- Comprehensive analysis and documentation of the evaluation exercises

Working Group 3: Applicability, Implementation and Practical Guidance

- Review of stakeholders requirements
- Compilation of a Best Practice Guidelines document for the reliable application of dispersion simulation tools, translating scientific output of the Action into stakeholder support

Following, the main results achieved thus far and the on-going activity are presented.

13.2 The Action Progress Thus Far

The activities performed thus far are documented in scientific reports available on the Action's website: http://www.elizas.eu. A short description of the main outcomes and some illustrative results are presented hereafter.

13.2.1 The Background Document

The first outcome of the Action was a state-of-the-art report [2] that apart from a general overview its focus is on specific problems related to dispersion modelling for emergency planning and response. The main topics included in the report are:

Analysis and assessment of the applicability of an ADM into ERTs, of the specific needs and possible improvements connected to the expected timely response, and of the reliability of current local-scale modelling techniques;

Definition of the concept of threats, description of threat scenarios, source terms of concern, critical and challenging situations for the different communities involved in local-scale emergency response;

Review of the different modelling approaches and tools, the limitations of both simple and advanced models of emergency response systems from a current perspective;

Discussion of the particular challenges for contaminant dispersion modelling applied to the local-scale and of the needs for future model development;

Analysis of the present evaluation process for local-scale dispersion models, in particular when dealing with the uncertainties related to the application of models in emergency response;

Discussion of the specific requirements and datasets to pursue the quality assurance of local-scale models and of the related evaluation methodologies, including preliminary guidelines;

Discussion of the importance of the interaction between scientists and model developers with end-users, stakeholders and decision makers;

Outline of practical constraints, regulations and legal issues.

13.2.2 The Inventory of Available Datasets

A database was design to include datasets suitable for validation of dispersion models that can be integrated in emergency response systems [8]. Since datasets suited for emergency response models are rare, mostly atmospheric dispersion experiments are included in the database. The classification of databases is based on the Action's main goals, which refer to (1) Accidental (even when intentional) releases, and (2) Built-up environments. The limitations of using data to validate models in emergency response assessment are discussed as part of this document. In a second version, some datasets that better comply these two requirements were selected. Among them, three were chosen as case studies for the Action modelling and evaluation activity: the Michelstadt exercise (wind tunnel), presented in the following, a combined wind tunnel and field campaign with continuous and puff releases resembling the Hamburg harbor and a real industrial accident occurred in a European Country. The model evaluation and inter-comparison will be completed within the year 2014.

13.2.3 The Inventory of Emergency Modelling Tools

A catalogue of state-of-the-art emergency response tools and a model inventory for airborne hazards from accidental/deliberate releases in complex urban and industrial areas was prepared: the Emergency Response Models and Tools Inventory Database Tool (ERMIDT) [7]. It collects detailed information on existing models and ERT currently applied in the context of the Action, developed for local-scale incident scenarios. The structure of the catalogue enables an efficient access to the required information: type of application, type of computational approaches and models integrated, aspects of hazards and scenarios addressed, physical background, input data demands, model outputs, computational demands and information on model application/use, verification or related performance measures. The inventory is intended to support model-specific guidance regarding an efficient and reliable use of different models and tools.

13.2.4 The 'Michelstadt' Modelling Exercise

The first modelling exercise, "Michelstadt", is based on data gathered in a wind-tunnel flow and dispersion experiment performed in the WOTAN atmospheric boundary layer wind tunnel at the Environmental Wind Tunnel Laboratory in Hamburg [3]. The measurements were carried out in an idealized Central-European urban environment model, named as Michelstadt. Five point sources were used non-simultaneously in continuous and short-term release mode, and two wind directions were investigated. Model simulations and an intercomparison for continuous and

Table 13.1 Types of models used to simulate Michelstadt test case

Dispersion modelling method	Number of different models	Computational time
Gaussian (2 with building parameterization)	7	1–5 min
Lagrangian	5	2 min–5 h
CFD (8 RANS; 3 LES; 1 RANS-Lagrangian)	10 (6 models)	2 h–4 days

puff releases were performed, aiming at identifying the key aspects and possible problems arising in applications to emergency response cases. The different modelling approaches are summarized in Table 13.1.

Blind and non-blind cases were simulated. Flow and concentration data were provided for the non-blind test case, while in the blind test only the minimum flow information was provided to the modellers. Initial results obtained with models of Type II [9] show relative good agreement between observed and predicted mean concentration for a continuous release in non-blind simulations. However, a lower score is obtained (as expected) for simulations of the puff release in the blind case.

A statistical analysis of the results was reported in [1]. The metrics considered in the analysis was the normalized mean squared error (NMSE) of the mean concentration in the continuous release and the mean dosage in puff releases cases, both in the non-blind and blind test cases. The acceptance criterion in built environment requires that NMSE < 6 [5]. Most models give performances within the acceptance limit for the non-blind case while there is a setback for the blind case; a certain variability in the performances of different models was also reported.

13.2.5 The Data Comparison Tool

An ad-hoc tool for comparing physical measurements and results of numerical simulations was developed in Python and it has been already applied to the Michelstadt modelling exercise [6]. The main features of the tool are:

- "User friendly" as well as "Advanced user" modes;
- generic and flexible, applicable to models of any complexity, with different outputs (object oriented programming);
- built in a modular way in order to facilitate the easy inclusion of additional metrics, plots, etc;
- developed to be used both under Linux and Windows;
- includes all modules necessary to produce the required outputs (metrics, plots).

On the basis of the first test (Michelstadt exercise), further statistics and graphical processing that can be useful specifically for model evaluation in continuous or puff releases were identified and are being implemented in the tool to be used in next Action's modelling exercises.

13.2.6 The End-Users and Stakeholders Questionnaires

A questionnaire surveying the present tools used by end-users and stakeholders was elaborated and distributed to investigate their needs and requirements related to the modelling suites. Not surprisingly, according to questionnaires, most of the responsible agencies use simple approaches with minimal meteorological input and no consideration of buildings. Only few of them reported using sophisticated models in combination to mesoscale meteorological models. The opinions of stakeholders towards uncertainties in the modelling process were also analyzed. The preferences expressed by stakeholders and end-users with respect to the type of model outputs from an ADM are: hazardous areas on maps, hazardous distances, concentration values and confidence intervals. The stakeholders' expectation from the use of ADM can be shortly summarized as: emergency models have to be simple, robust and fast, they should provide user-friendly interface with on-line help and supply potential damage zones on "google maps"-like as output.

As a follow-up, more in-depth interviews were conducted and analysed [4]. The general opinion expressed about the usefulness of ERTs goes from 'good' to 'essential'. The practical limitations in using more complex tools in ERTs are a major concern for end-users and decision makers. Documenting the limitations of different local-scale emergency response methodologies by assessing the actual uncertainty of model results is recognized as an important issue as well.

13.3 The Ongoing Activity

13.3.1 The Catalogue of Threats and Challenges

The scope of the catalogue is to collect, document and characterize typical and relevant local-scale threats related to releases of toxics agents in populated areas. The goal of the catalogue is to guide the model development towards the present and future needs of emergency response management. The document provides a description of conditions leading to release of hazardous materials, which might impact negatively exposed humans' health and safety. Then, main topics pertinent to the consequences analysis are addressed.

13.3.2 The Model Evaluation Protocol

The scope of this document is to review model evaluation procedures for the validation of dispersion models, which can be applied in cases of accidental or deliberate releases of airborne hazards in urban areas. A task-oriented model evaluation protocol is proposed, starting from the requirement to introduce an evaluation procedure applicable during all three distinct phases in emergency response: pre-accidental analysis and planning; predictions during an actual emergency and post-accidental analysis. The protocol is tested and further improved in the course of the modelling exercises planned in the Action.

13.3.3 The Best Practice Guidelines

The objective of the document is to provide guidance on the application of ADMs in emergency response in order to minimize the uncertainty in simulation results and increase their reliability. The document provides comprehensive information on usability, pros and cons as well as challenges and limitations of different modelling approaches. The document highlights the location of the ADM within the chain of assessment of an ERT, and identifies the model results that can be used in an operational approach. The document suggests possible approaches to application of models under conditions typical to emergency events: scarce or lack of input data, field measurements ingestion, etc. These Best Practices are provided for the different phases of emergency: pre-accidental/planning, during emergency and post-accidental.

References

1. Baumann-Stanzer K, Leitl B, Trini Castelli S, Milliez M, Rau G, all COST ES1006 Members (2014) Evaluation of local-scale models for accidental releases in built environments – results of the "Michelstadt exercise" in COST Action ES1006. In: Proceedings of the 16th international conference on harmonisation within atmospheric dispersion modelling for regulatory purposes, Varna, Bulgaria (in press)
2. COST ES1006 – Background and Justification Document, COST Action ES1006, May 2012. http://www.elizas.eu/images/Publications/costes1006_d1_final_low-res.pdf
3. Fischer R, Bastigkeit I, Leitl B, Schatzmann M (2010) Generation of spatio-temporally high resolved datasets for the validation of LES-models simulating flow and dispersion phenomena within the lower atmospheric boundary layer. In: Proceedings of the 5th international symposium on Computational Wind Engineering (CWE2010), Chapel Hill, North Carolina, USA
4. Gariazzo C, Leitl B, Trini Castelli S, Baumann-Stanzer K, Reisin TG, Barmpas F, Tinarelli G, Milliez CM, Bemporad E, Armand P (2014) COST Action ES1006. Evaluation, improvement and guidance of local-scale emergency prediction and response tools for airborne hazards in built environments: ongoing activities, experiments and recent results. Chem Eng Trans 36:529–534, 2014-CISAP6
5. Hanna S, Chang J (2013) Acceptance criteria for urban dispersion model evaluation. Meteorog Atmos Phys 116:133–146
6. Milliez M, Hellsten A, all COST ES1006 members (2014) Evaluation of dispersion models for improvement and guidance for the use of local-scale emergency response tools; the "Michelstadt" modelling exercise – COST action ES1006. In: Proceedings of the 6th international symposium on Computational Wind Engineering (CWE2014), Hamburg, Germany, 8 p

7. Tavares R, Baumann-Stanzer K, Leitl B, Trini Castelli S, all COST ES1006 Members (2014) ERMIDT – the Emergency Response Models and Computational Tools Inventory Database developed in the frame of in COST Action ES1006. In: Proceedings of the 16th international conference on harmonisation within atmospheric dispersion modelling for regulatory purposes, Varna, Bulgaria (in press)
8. Tsiouri V, Trini Castelli S, Andronopoulos S (2013) Qualified datasets for benchmarking local-scale emergency response models for the purpose of disaster management. In: Proceedings of the 15th international conference on harmonisation within atmospheric dispersion modelling for regulatory purposes, Madrid, Spain, pp 27–32
9. Trini Castelli S, Leitl B, Baumann-Stanzer K, Reisin TG, Armand P, Barmpas F, Balczo M, Andronopoulos S, Jurcakova K, all COST ES1006 Members (2014) Updates on COST Action ES1006. Evaluation, improvement and guidance for the use of local-scale emergency prediction and response tools for airborne hazards in built environments. In: Proceedings of the 16th international conference on harmonisation within atmospheric dispersion modelling for regulatory purposes, Varna, Bulgaria (in press)

Part III
Prevention and Mitigation

Chapter 14
Focusing on the Threats to the Detriment of the Vulnerabilities: A Vulnerability Assessor's Perspective

Roger G. Johnston

Abstract Security and emergency response usually fail due to a failure to identify and mitigate vulnerabilities, rather than a failure to envision threats. Unfortunately, however, threats usually receive more attention than vulnerabilities. Design Basis Threat and layered security exacerbate the problem. There are a number of aspects of nuclear security that require considerably more attention to the vulnerabilities.

14.1 Overlooking/Under-Emphasizing Vulnerabilities

Effective preparation and risk management for any kind of security application requires not just focusing on the threats, but also understanding and mitigating the vulnerabilities. In my experience, it is very common for threat assessments (TAs) to receive far more emphasis than vulnerability assessments (VAs), often to the serious detriment of security.

It is important to be clear about what constitutes TAs and VAs, and about the difference between threats and vulnerabilities. There often seems to be a great deal of confusion about these issues.

Definitions vary, but it is useful to define a **threat** as who might attack, why, when, and how, and with what resources and probability. A **threat assessment** is an attempt to identify the threats. In contrast, a **vulnerability** is a security weakness that could be exploited by the threats to cause undesirable consequences. A **vulnerability assessment** involves discovering and demonstrating these weaknesses and ways to defeat a security device, system, or program. An effective VA also often includes suggestions for counter-measures and security improvements.

R.G. Johnston (✉)
Right Brain Sekurity, 5 Navajo Court, Oswego, IL 60543, USA
e-mail: roger@rbsekurity.com

S. Apikyan, D. Diamond (eds.), *Nuclear Terrorism and National Preparedness*, NATO Science for Peace and Security Series B: Physics and Biophysics,
DOI 10.1007/978-94-017-9891-4_14

For example, one possible threat to a nuclear program might be terrorists wanting to steal nuclear material by using outsiders, insiders, or both. Possible vulnerabilities might include a nuclear facility that is poorly designed for security, intrusion detectors that are easy to tamper with, poor insider threat mitigation, absent or ineffective training about social engineering, and a non-secure chain of custody for procured hardware and software.

The Vulnerability Assessment Team at Argonne National Laboratory has found that some things frequently get confused with vulnerabilities. These include the assets to be protected (e.g., nuclear material), threats, delay paths, facility or building features (e.g., gates or nearby roads), attack scenarios, or safety issues [1, 2]. Although these things may have important implications for security vulnerabilities, they are not the actual vulnerabilities themselves. Another common problem is that so-called "performance testing" is often thought of as a VA in itself when, in fact, it is only a small part of a comprehensive VA.

I have often observed confusion over the purpose of a VA. The purpose of a VA in my view is to (1) improve security and (2) serve as one of the inputs to Modern Risk Management. (See Fig. 14.1.) It is NOT the purpose of a VA to do any of the following:

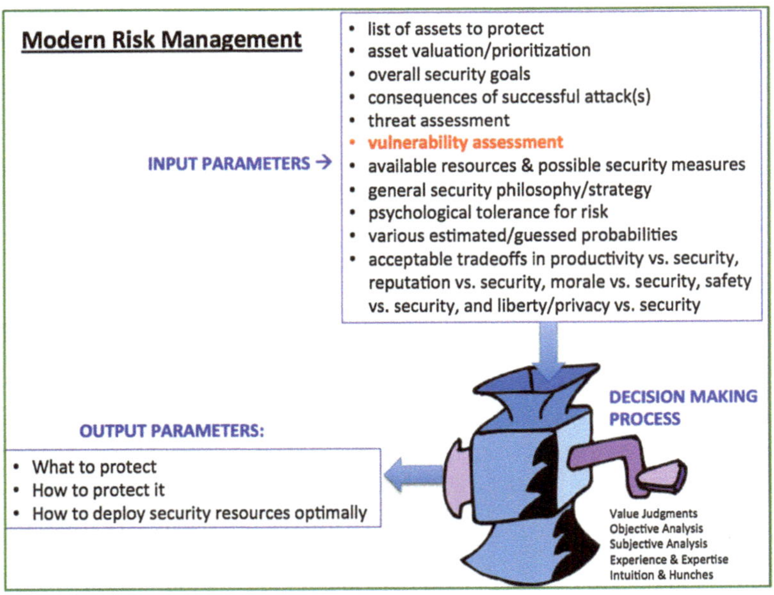

Fig. 14.1 Modern Risk Management involves a number of inputs, including information obtained from threat assessments (TAs) and vulnerability assessments (VAs). All these inputs are used in the decision-making process, along with value judgments, objective and subjective analysis, experience, expertise, intuition, and hunches. As a result of this decision making process, we determine what to protect, how to protect it, and how to field our security resources for optimum security

- Pass a test
- "Validate"
- Serve as auditing
- Generate metrics
- Do a safety analysis
- Justify the status quo
- Praise or accuse anybody
- Check against regulations
- Check against some standard
- Engender warm and happy feelings
- Claim that there are no vulnerabilities
- Determine who gets salary increases
- "Test" security or do performance testing
- Rationalize the research and development
- Apply a mindless, bureaucratic stamp of approval
- Claim there has never been a loss of nuclear material
- Endorse a security product or program, or certify it as "good" or "ready for use"

People sometimes talk about a security device, system, or program "passing a vulnerability assessment". This is a nonsensical statement; it certainly should not be construed to mean there are zero vulnerabilities. All security devices, systems, and programs have vulnerabilities—usually in very large numbers—and not all these vulnerabilities can ever be found. What people probably mean by this statement is that a VA was conducted and then *somebody* made a value-judgment (based on a variety of factors) that the existing vulnerabilities can be tolerated. Their conclusion, however, is not the responsibility or prerogative of a VA or of vulnerability assessors.

Another problem is that security is often analyzed in terms of safety. While safety considerations can reveal important things about what needs to be protected and how adversaries might attack, safety is not a good basis for planning security. The presence of a nefarious, purposeful, adaptable adversary who attacks intelligently at the weakest points is missing in safety applications. Thus, safety and security are not the same kind of problem and should not be analyzed in the same way.

A number of different security techniques—while still potentially useful—frequently get confused with VAs. These include security surveys (walking around with a checklist), security audits (checking if the rules are being followed), feature analyses, threat assessments, Design Basis Threat (see the discussion below), fault or event tree analyses (from safety engineering; often very problematic for security analysis), the Delphi Method (a technique for getting a decision from a panel of experts), software assessment tools, the CARVER Method (often used by the U.S. Department of Defense and law enforcement), and Modern Risk Management.

14.2 Examples

Here are a few of many possible examples of serious security incidents or natural disasters for which the threat was fairly well understood but the vulnerabilities were either not understood or not dealt with. Poor preparation or security resulted. Several of these are examples of Michener's Maxim: We are never prepared for what we expect.

Hurricane Katrina, 2005: The threat and the relatively high probability of a severe hurricane eventually hitting New Orleans was well known, but the city's vulnerabilities were unknown or ignored, as were the weaknesses in the emergency/disaster response systems [3].

Breach of the Y-12 nuclear facility by an 82-year-old nun and two other protesters, 2012: The threat represented by protestors and the risk of trespassing were presumably well understood given their long history at U.S. nuclear facilities, yet the alarm and camera systems at Y-12 repeatedly malfunctioned or were broken, and security guards responded poorly to indications of intrusion [4].

Target stores credit card hack, 2013: While the threat of stealing credit card information and potential attack methods were apparently well understood, Target reportedly ignored information from its own $1.6 million security system that data breaches had occurred [5].

White House fence jumper, 2014: A man jumped the fence at the White House in Washington D.C. and sprinted across the lawn, entering the White House through an unlocked front door and penetrating a distance into the building. A number of security features failed to be deployed or to work properly [6]. Although the threat was well understood given the long history of White House intruders [7], including fence jumpers, the vulnerabilities were not adequately recognized or dealt with.

14.3 Why Threats Are More Popular

Given that critical security often fails due to vulnerabilities, why is it that threats usually receive more attention? There are probably a number of reasons. First of all, threats are easier to anticipate. Most security programs face a relatively small number. There will, however, typically be several orders of magnitude more vulnerabilities for any non-trivial security application. Moreover, while many threats are obvious based solely on past history, common sense, and global issues, this is often not true for vulnerabilities. A thorough understanding of vulnerabilities requires a comprehensive understanding of the exact, local details of the security program or application. (The devil is always in the local details.) With threats, on the other hand, security experts with no knowledge of your particular security application can still do a competent job of identifying many of your threats.

Another factor is that vulnerabilities often take more imagination to recognize than threats, and require "thinking like the bad guy". Imagination and creativity are characteristics that are often not the strong suite of bureaucracies, military organizations, security professionals, and large nuclear programs. And security professionals and engineers—being the good guys—often have a very different mindset than do the nefarious adversaries. Moreover, vulnerabilities, much more so than threats, are often best revealed by using techniques that are not formalistic, quantitative, reproducible, or objective [8]. Governments and large organizations, however, typically do not like to rely on such subjective methods for something as important as nuclear security. Given the dearth of research-based practice when it comes to nuclear security, and given the fact that subjective methodologies are often the best for seeing into the future [9], insisting on only formalistic, objective analysis seems imprudent and risky.

Cognitive dissonance is also an issue. Whereas few security programs claim to face zero threats, many deny (or want to deny) that they have significant vulnerabilities. Identifying vulnerabilities is often taken to be an implied criticism of a security program, even though vulnerabilities are always present in large numbers. Citing threats, however, is not usually taken as criticism. Whereas "shooting the messenger" is rarely a problem with TAs, it is a major occupational hazard for vulnerability assessors.

TAs tend to be reactive to past security incidents. VAs are usually more proactive, and try to predict what has not yet occurred. Being proactive with security is always more challenging, uncertain, and politically dangerous than being reactive.

Finally, TAs are typically more reproducible than VAs, and this can be quite reassuring. Whereas different threat analysts often see the same or similar threats, different vulnerability assessors (or the same vulnerability assessors at different times) often find quite disparate vulnerabilities or miss critical ones.

14.4 Vulnerabilities Trump Threats

Modern Risk Management requires a thorough understanding of vulnerabilities as well as threats. A number of other factors need to be well understood, as well, as shown in Fig. 14.1.

While it is certainly true that effective security must be tailored to meet the actual threats, and that both threats and vulnerabilities are important, a sophisticated understanding of vulnerabilities is often better than a sophisticated understanding of threats. Imagine a situation in which we have a 100 % accurate understanding of the threats we face, but are totally clueless about our vulnerabilities. There is little chance our security will be successful because the known threats will be able to find and exploit our myriad weaknesses.

On the other hand, imagine that we have a sophisticated understanding of our vulnerabilities and have mitigated or eliminated those that we can, but we don't have the slightest idea who might attack when, or why, or with what probability. This is not an ideal situation, but our security might still work because we have attenuated or eliminated many of the serious vulnerabilities that the (unknown) attackers could exploit. It would certainly be better to understand the threats so that we could field resources more appropriately. Nevertheless, the understanding we have of our vulnerabilities still goes a long way toward countering the adversary.

While both effective TAs and effective VAs are needed for good security, there are other factors that can give more "bang for the buck" for VAs than for TAs. TAs are speculations about groups and people who may or may not exist, as well as their goals, motivations, and resources. Vulnerabilities, on the other hand, are not theoretical. They are right in front of you—if you open your eyes and mind—and they are often testable. It is also often possible to test countermeasures to the vulnerabilities.

In my experience, vulnerabilities are frequently easy and inexpensive to mitigate or eliminate once they are acknowledged, whereas it is usually difficult or impossible to do the same to a threat. The problem with ignoring vulnerabilities that we think nobody can exploit is that we usually are wrong about that.

Finally, an understanding of vulnerabilities trumps an understanding of threats because bad guys don't do TAs. What they do when they are contemplating or planning an attack is closer to conducting a VA than a TA because they typically want to attack at the weakest points. So if we want to predict what the bad guys are likely to do, we would be wise to try to think like them, get inside their heads, and examine our security in the context of vulnerabilities.

14.5 Design Basis Threat and Layered Security

Design Basis Threat (DBT) is a technique often used for securing nuclear facilities and material. In practice, DBT often ignores or under examines security vulnerabilities. In my view, it also typically suffers from an obsession with force-on-force attacks, fails to focus adequately on sabotage issues, and incorrectly ignores threats judged (often mistakenly) to be of low probability. DBT typically ignores or minimizes issues associated with the insider threat, and frequently ignores the issues of mitigating employee disgruntlement. This is particularly unfortunate because in every case of nuclear theft where the details are known, insiders were involved [4].

The common use of DBT as a way to "test" one's security is, I believe, particularly unfortunate in that it represents nonsensical circular logic. The security is defined with a speculative mental model of the threat, and then the actual security implementations are "evaluated" based on the assumed accuracy of the very model that defined the security problem in the first place.

DBT is often associated with layered security ("security in depth"). Although layered security is often necessary, it frequently fails spectacularly and stupidly, as in the above 4 examples just discussed. Reliance on layered security seems to reduce critical and imaginative thinking about security vulnerabilities. Certainly both DBT and layered security are poor substitutes for the Modern Risk Management shown in Fig. 14.1.

14.6 Areas Needing Significant Improvement

As a vulnerability assessor, I see a number of areas in nuclear security that need significant improvement. We need far more research-based practice. Little of what is done in the field of nuclear security is based on any significant research. For example, the two-person rule originally arose from safety considerations. It is now the basis of a lot of nuclear security, yet there has been very little research to show if it has merit or how to best implement it. In fact, recent psychological research may even call the rule into question [10].

We also need improvements, better techniques, and more VAs for insider threat mitigation, especially disgruntlement mitigation for employees, contractors, consultants, and vendors. There are many tools available for this purpose, but they are rarely exploited for nuclear applications. There are also techniques for reducing security guard turnover that are rarely deployed [11].

The chain of custody for procuring nuclear hardware and software must be made much more secure, starting right at the factory. In my experience, only 15-s of access is all that is typically required for a nefarious adversary to compromise a security device. Testing whether a security device or system behaves normally is of little value for detecting such tampering. Few security and safeguards devices have effective tamper detection features.

There are numerous potential, but largely unexploited psychological, managerial, and organizational countermeasures to insider threat, security theater, cognitive dissonance, groupthink, perceptual blindness, sleight-of-hand/misdirection, and poor supervision of front-line security officers. We must also fight the false but ever present idea of "security by obscurity". Somewhat counter-intuitively, security is best when it is largely transparent. Openness allows review, criticism, accountability, and buy-in. Long-term secrets cannot be kept reliably, and are not conducive to achieving effective security or to implementing security improvements.

We need to stop confusing safety and security. We need to stop letting safety experts plan security measures, and we need to stop relying so much on techniques that are primarily safety-based. We need to better incorporate the concept of continuous improvement for physical security and nuclear safeguards. We need to embrace the new security paradigms [12].

Better tags and tamper-indicating seals, and better use protocols are possible and necessary [13, 14]. We need more secure real-time monitoring and transport of radiological and nuclear materials, more secure access control devices, improved hardware protection for secret passwords and secret keys, and better (simple) security for sealed radiological sources. We need to avoid thinking that we can apply security to devices, buildings, or facilities at the last minute, in a kind of "Band-Aid" approach. Effective security requires early and iterative VAs, beginning at the design stage.

Another common problem involves confusing inventory functions with security functions. This is much of the reason why global positioning systems (GPS), radio frequency identification tags (RFIDs), and many nuclear control and accountability (MC&A) programs provide poor security. They are not fundamentally designed to counter deliberate spoofing and surreptitious attacks.

14.7 Conclusion

Effective security and preparation for nuclear incidents requires much more frequent, comprehensive, and effective vulnerability assessments. There are many possible ways to improve nuclear security if the problem is viewed from the perspective of vulnerabilities, vulnerability assessors, and nefarious adversaries.

Acknowledgments and Disclaimer The views expressed here are those of the author and should not necessarily be ascribed to Argonne National Laboratory or the United States Department of Energy. Jon Warner provided important ideas and expertise.

References

1. Johnston RG (2010) Being vulnerable to the threat of confusing threats with vulnerabilities. J Phys Secur 4(2):30–34
2. Johnston RG, Warner JS (2013) Debunking vulnerability assessment myths. SecurityInfoWatch, August 6 & 13, 2013, Part 1: http://www.securityinfowatch.com/article/11078830/experts-discuss-commonly-held-misconceptions-about-vulnerability-assessments. Part 2: http://www.securityinfowatch.com/article/11108983/experts-discuss-the-characteristics-of-good-vulnerability-assessors
3. American Society of Civil Engineers (2007) The New Orleans Hurricane protection system. What went wrong and why? http://www.asce.org/uploadedfiles/publications/asce_news/2009/04_april/erpreport.pdf
4. Bunn M, Sagan SD (2014) A worst practices guide to insider threats: lessons from past mistakes. American Academy of Arts and Sciences. http://cisac.fsi.stanford.edu/sites/default/files/insiderThreats.pdf
5. SlashGear (2014) Target reportedly ignored credit card hack warnings, March 14, 2014. http://www.slashgear.com/target-reportedly-ignored-credit-card-hack-warnings-14320673/
6. Leonnig CD (2014) White House fence-jumper made it far deeper into building than previously known. Washington Post, September 29, 2014
7. Wikipedia, "White House Intruders". http://en.wikipedia.org/wiki/White_House_intruders
8. Johnston R, Folk C, Warner J (2014) Fearing NORQ, and why doing so is bad for security. Homel Secur Today 11(4):39–41. http://www.nxtbook.com/nxtbooks/kmd/hst_20140607/#/40
9. Harford T (2014) How to see into the future. Financial Times, September 6, 2014. http://www.ft.com/cms/s/2/3950604a-33bc-11e4-ba62-00144feabdc0.html#axzz3GhgHv4aA
10. Wiltermuth SS (2011) Cheating more when the spoils are split. Organ Behav Hum Decis Process 115(2):157–168
11. Bitzer E (2006) Strategies for cutting turnover. Secur Manag 50(5):88–94
12. Johnston RG (2010) Changing security paradigms. J Phys Secur 4(2):35–47
13. Johnston RG (2005) The 'Anti-Evidence' approach to tamper-detection. Packag Transp Storage Secur Radioact Mater 16:135–143
14. Johnston RG, Warner JS (2012) How to choose and use seals. Army Sustain 44(4):54–58. http://www.almc.army.mil/alog/issues/JulAug12/browse.html

Chapter 15
Preventing, Controlling and Mitigating Radiological Consequences

Marcos Antonio Do Amaral, Alfredo Lopes Ferreira Filho, and Josilto De Aquino

Abstract The Brazilian Radiological Protection Society – SBPR is the entity that aggregates almost all the radiological protection experts in Brazil. Due this integration of expertise, many of the strategies regarding preparedness and mitigation of accidents on nuclear or radiological sources and preventions against tragic events made by men, are prepared according the directions of the members of SBPR, working on utilities, governmental and or educational institutes.

In this context, the emergencies are handled following the standards and regulation defined by the Brazilian National Nuclear Energy Commission – CNEN. The CNEN also adopts the guidance given by the IAEA GSG-2 – Criteria for Use in Preparedness and Response for a Nuclear or Radiological Emergency and, for reactors, the Tec. Doc. 955 – Generic Assessment Procedures for Determining Protective Actions During a Reactor Accident.

This paper is according the point of view of the SBPR, with a short description of a desirable organization and relationships in order to enable a city, state or country to be prepared and to handle a possible nuclear accident or even a radiological accident happening on any application area.

Within the comparison between the different point of views, presented by different expert organizations or authorities, the Society considers that giving an additional perspective may contribute for the overall improvement of the emergency plans in many places, thus contributing for the safety and protection over the world.

M.A. Do Amaral (✉) • A.L.F. Filho • J. De Aquino
Brazilian Radiological Protection Society, Rio de Janeiro, Brazil
e-mail: amaral1@eletronuclear.gov.br

S. Apikyan, D. Diamond (eds.), *Nuclear Terrorism and National Preparedness*, NATO Science for Peace and Security Series B: Physics and Biophysics, DOI 10.1007/978-94-017-9891-4_15

15.1 Introduction

The increased technology enhancements of the humankind in our days enables the increase on the destruction power of possible malicious actions made by men. In parallel, the present climatic changes with indicators pointing for future increases on the intensity of the changes, contribute to increase the risks for natural catastrophes over the planet.

In this context, fundamental is to improve defence strategies in order to face the increased threats, specially radiological or nuclear threats, encompassing both adequate preparedness and response to achieve the best control in face of the emergencies [1].

The response to a nuclear or radiological emergency expects to involve several organizations, whose functions would be the same for a nuclear or radiological emergency as for a conventional emergency, only the response to a nuclear or radiological emergency may also involve highly specialized agencies and technical experts. The effective response to a nuclear or radiological emergency must be well co-ordinated and appropriately integrated with those for a conventional emergency. The misunderstandings regarding health effects from radiation are contributors for non-proportional decisions taken in face of the real detriment, challenging the pre-planned actions to control and mitigation of the emergency.

15.2 Background

The knowledge regarding previous accidents and or threats is crucial for the best response by the emergency teams [1]. The decisions taken, difficulties with supplies and consumables, transportations, communication, number of victims, escape routes, clothes and issues from respirator protection devices etc, are valuable information for enabling the responders to apply the better response.

Beyond the information, training for practical situations, based on the above and other simulated scenarios, may be the key for mitigating or annulling the accident consequences.

Adequate resources, both material and human, should be available and functional, and strategically located for promptly use or delivery for use. Thus, the emergency response centres ideally should be closely located to the affected area, but far enough to avoid the need of direct protection measures for the emergency responders working inside the centres.

All those items above, between others important items, are to be addressed by an adequate organization and management, clearly specified on the emergency plan and procedures.

15.3 Plans and Procedures

The Emergency Response Plan and related Emergency Response Procedures are the fundamental documents to establish the organization for management of any nuclear or radiological emergency. The Plan and Procedures should cover the response on-site and offsite, for the different disciplines involved, and periodically tested under simulation scenarios [2–4].

15.4 Monitoring for Prevention

The best way for preventing any emergency is to monitoring the installations and vicinities, by using radiation detectors, periodical surveys, trend evaluation by using indicators, self-assessments, external audits, inspection of authorities, effective supervision, adequate safety culture, reliable engineering controls and full adherence to the administrative controls. Beyond the prevention, in case of any emergency happening, the monitoring tools together with the deep defence barriers will make the difference for the success or disaster in face of the emergency.

15.5 Attributions and Responsibilities

The organizational structure and assignments includes the roles and responsibilities for all the players in the emergency plan, like licensees, emergency workers and emergency responders, local and national agencies and other organizations [1].

In some situations, external experts would be important for dealing with the emergency. For such situations, the emergency procedures should specify the authority to assign a person as emergency responder, also defining the criteria for required qualifications for each category or discipline needed on the specific emergency.

15.6 General Provisions

The anticipation of external support and defined provisions are important to host, feed and transport the external support organizations to integrate them with the emergency response team.

A standard emergency classification and action level scale, based on the INIS and deeply detailed on the procedures, may easier the understanding the consequences or potential consequences of an evolution of an accident, enabling the adequate and timely response.

Communications are essential. They should be promptly available and by multiple pathways, both for public and internally, comprising telephones, social media, professional networks, newspaper, radio, TV etc.

Communications experts for each media should be allocated as members of the emergency organization, and following the directions from the emergency coordinator. Periodic and timely communications informing the accident evolutions are a key action to demonstrate control and to inspire trust, also reducing the population anxiety, both affected and non-directly affected by the emergency.

Assessment of dose for the radiological releases consequences, by using direct and indirect techniques, should be in place. The type of the radiation (beta, alpha, neutron, electromagnetic waves), its chemical composition (noble gases, halogens, metals) and the physical form of the release (aerosol, gaseous, liquid, solid debris) should be known in order to estimate the consequences. This is done by using radiation monitors, both fixed and mobile, air samples, or may be previously modelled and estimated by using computer codes. The person in charge for performing the estimates of radiological consequences shall be formally qualified and assigned for this function, reporting directly to the radiological protection manager or to the general emergency coordinator.

Required protective actions like evacuation, sheltering, thyroid blocking with potassium iodide, prevention from consuming contaminated food and water, established according reference levels, should be taken in a timely manner, according the degree of severity of the emergency. In the same way, the control over radiological exposures in emergency shall be in place, with the reference levels for emergency dose according the threats to be avoided and the authority approval for emergency exposures formally assigned at the competent level.

Medical care is expected to support the situation where internal contamination or elevated exposure could cause injuries to the affected individuals.

Emergency coordination centres should be strategically located, and personnel adequately trained to fulfil the demands for controlling and mitigating the emergency. Such centres may be located, if possible, inside the affected site if there is a warranty of habitability and protection resources to fight the emergency. However, is fundamental to have an off-site emergency centre, providing extra personnel to respond to the accident and, in case of required relocation of the onsite emergency centre, receiving experts from that centre and providing the necessary resources for their work. Also, is expected the offsite emergency centre holds the coordination between the licensee and government officials, civil defence forces, fire fighters and other organizations playing some role at the emergency.

At the emergency centres, it should be available the information regarding radiation levels, assessment of doses both for personnel and from radiological releases, meteorological parameters, operability of the affected facility systems, survey maps, injured personnel, security issues, decontamination and repairs needed, and operational records.

15.7 Protecting the Emergency Workers and Emergency Responders

Emergency workers and or personnel not employed by the affected installation, like firefighters, physicians, rescuers etc, must to receive at least passive dosimeters and should be monitored and protected while accomplishing their missions. For the best monitoring, the exposed personnel should wear direct reading dosimeters or electronic dosimeters, according their availability. More than this, it is expected they are well trained and medically tested before performing duties as emergency worker, also being aware about the risks associated with their job, especially regarding the emergency in hands.

15.8 Radiation Protection Officer Actions

A fundamental key is the event classification according the radiological risks, in order to select the appropriate protective actions for the public and for the workers.

The access control is a fundamental part of the emergency control and mitigation. Only classified personnel can enter inside the emergency controlled area, and participating of the mitigation efforts. Non-classified people also represents a matter of concern, as they demand resources and attention for giving them the necessary protection inside the affected area, instead directing those efforts to the ones really requiring such protection. Thus, is very important to exclude the non-necessary personnel from the affected area, establishing the secured perimeter in order to permit the focus and attention from the emergency workers and responders to the emergency control and mitigation.

15.9 Social Communication

During an emergency, a strong enemy is the fear that both people of the public and conventional workers exhibits from the radiation. This fear sometimes is increased by misinformation coming from pseudo experts, some them from academy, social media and other public personalities. The way to attenuate those fears are using clear and concise communication from the operators, authorities and renowned experts in the area, timely and periodically. Such communication should use all the available Medias, including internet applications like Twitter, Facebook, Linkedin, Whatsapp etc.

On places where nuclear energy is such an evil, mainly caused by misinformation from groups ideologically aligned, more important yet is the correct use of the communication techniques. In this sense, very important is the availability of mass media communication experts, giving advice, supporting the production of bulletins and interviews with the emergency coordinators in order to give the better idea about the emergency, and maintaining the population attempt and cooperative in case other protection actions are need to be taken preventively.

15.10 Mitigation

Many techniques for mitigating the emergency consequences are possible of use. When removing the affected population sometimes may not be the best alternative, because the possible high costs involved related to the avertable dose to obtain, sheltering is an important approach. After, confining the source term, shielding the sources, decontaminating the affected areas and habitable spaces are logical steps to perform, according possible. Clear information regarding to the radiation levels are critical points to enable the emergency workers and emergency responders for the adequate self-protection, as they may promptly act to mitigate any radiological consequences for others on the affected space. Prompt bio analysis should be available to assess doses for affected individuals or potentially affected, contributing to reduce the anxiety degree for both public and emergency workers and responders, also Identifying possible injured persons by the radiation and directing them for adequate medical cares [1].

References

1. Occupational Radiation Protection in Severe Accident Management EG-SAM Interim Report (2014) ISOE IAEA/NEA-OECD - Paris, France
2. GS-R-2 (2002) Preparedness and response for nuclear or radiological emergency. IAEA, Vienna
3. GS-G-2 (2011) Criteria for use in preparedness and response for nuclear or radiological emergency. IAEA, Vienna
4. IAEA TECDOC 955 (1997) Generic assessment procedures for determining protective actions during a reactor accident. IAEA, Vienna

Chapter 16
Nuclear Preparedness Through a National Pre-event Public Information Campaign

Robert M. Levin

Abstract Ventura County's pre-nuclear explosion public information campaign proved that education of the public on this important threat is possible without causing panic or criticism. Experts hoped that following the launch of this campaign FEMA would promote a similar message nationally. That has not happened. FEMA must either take this matter up on its own or the nuclear preparedness community must find a way to get it to fulfill its responsibility.

The recollections and opinions expressed in this narrative are the author's alone. They in no way are meant to express the views of Ventura County Department of Public Health.

Ventura County became interested in nuclear preparedness some 10 years ago. Why? Because we live next door to Los Angeles. Because Al Qaeda said they want to kill Americans, terrorize Americans and devastate our economy. Because only one kind of terrorist act does all three of those things. No infectious disease can do that—not even Ebola. We can control and limit contagion. A dirty bomb won't do that. It wouldn't dent our economy. Suicide bombers won't do all of those things. Improvised explosive devices planted along our roadways won't do it. Only a nuclear bomb, an improvised nuclear device (IND), will do that. And since Al Qaeda was making the threat, multiple simultaneous INDs could be anticipated to be the order of the day, their preferred method of operation.

At that time, I approached Commander David Tennessen from the Ventura County Sheriff's Department, the head of our county's Terrorism Working Group (TWG). I told him my concerns about the need for IND preparedness and my desire to make it happen in our county. I confess to being surprised by his response. He immediately turned over the bulk of the machinery of the TWG to IND preparedness planning. This continued for months. Years later I learned that some big event was

R.M. Levin (✉)
Ventura County Department of Public Health, Ventura, CA, USA
e-mail: Robert.Levin@ventura.org

S. Apikyan, D. Diamond (eds.), *Nuclear Terrorism and National Preparedness*, NATO Science for Peace and Security Series B: Physics and Biophysics,
DOI 10.1007/978-94-017-9891-4_16

scheduled to take place in the next months in Los Angeles and there had been some worrisome internet chatter picked up by law enforcement; terrorists might be planning on detonating a nuclear device.

We convened a gathering of local first responders which included law enforcement (including the Sheriff's Department, city chiefs of police, the California Highway Patrol (CHP) and the local FBI); fire (county and local cities as well as Federal from the military base at Port Hueneme); Public Health; HazMat (a part of fire); Emergency Medical Services (a part of Public Health); the local branch of the National Weather Service (NOAA); representatives from the business community; our Office of Emergency Services; city emergency planners; and information technology nerds.

Over 100 first responders met over a couple of years to address such questions as how many people were likely to flee into and through our county; should we block the roads and prevent Los Angeles residents from entering our county (some of these questions are going to appear naïve but this was 10 years ago and we were starting from scratch); should we wash off vehicles with hoses from fire trucks as they drive beneath highway overpasses; should we reverse some or all of the lanes of the southbound highway traffic, i.e., contraflow (Ventura County is north and west of Los Angeles); should we prevent by force if necessary people who want to go south <u>into</u> Los Angeles; if communications were down and we wanted to know which way the fallout cloud was heading, how would we find out?

The first responders made several assumptions: If an IND were detonated in Los Angeles, Ventura County would not only contend with the possibility or reality of fallout, but it would have the likelihood of 2 million additional people (its normal population is 840,000) entering the county from Los Angeles in 1 million cars, along with 200 trauma cases, 150 people with acute radiation syndrome, 600 with multiple injuries, 600 blinded either permanently or temporarily, and 7,500 people with potentially significant exposures to fallout. Shelter, food, water, clothing, restrooms, decontamination, medical services and psychological services would all be needed and would all be rapidly compromised.

Our group of 100 first responders decided that we should <u>not</u> block the roads and prevent Los Angeles residents from entering our county (in fact, we should welcome them); we should <u>not</u> wash off vehicles with fire hoses as they pass beneath highway overpasses (to do so would slow down evacuating traffic as people sought to have their cars get as thoroughly cleaned as possible); we should <u>not</u> implement contraflow on the southbound highway lanes (the CHP said that there are not enough traffic cones in all of California to block off any one of the highways coming out of L.A. I am still not convinced that some of the lanes shouldn't be reversed—besides, people are going to do it anyway whether the CHP likes it or not); and we should not prevent by force people who want to go <u>into</u> Los Angeles.

We concluded that Public Health should take the lead in the development of our county's plan because so many of the consequences of an IND detonation—exposure to prompt radiation, exposure to fallout, burns, trauma, blindness, psychological trauma—are health related. It was also true that the Health Officer was the one most interested in this and was willing to work on writing the plan on his off hours in the absence of any grant funding. This turned out to be a 243 page document entitled the Ventura County Nuclear Explosion Response Plan. It was designed to be both an educational document and an Annex to a Multi-Purpose Disaster Response Plan.

The Sheriff ordered his staff to develop a Law Enforcement Plan.

We developed a Plume Trackers Group. If communications were down from Los Angeles and we wanted to know which way the fallout cloud was heading, this group would acquire or create its own plume maps.

But ultimately we realized that we needed an educated public if we were going to have any significant impact on our nuclear preparedness. A Department of Homeland Security Study found that with the detonation of a 10-kt ground burst in Los Angeles, 280,000 lives would be saved if the public knew in advance to take protective shelter.

So we started planning a Pre-Nuclear Event Public Information Campaign. Its goal was to educate the public as to what to do if they ever learned that a terrorist's IND was detonated in what would probably be Los Angeles. Its primary objectives were to educate Ventura County residents how to survive, stay safe and help others in the event of a nuclear explosion. The secondary objectives were to promote an ongoing dialogue locally and to encourage other jurisdictions across the nation to engage in this dialogue as well.

Jonah Ansell from JAM Productions, volunteered to work with me pro bono to develop a strategy for the campaign.

We participated in national calls and conferences; one that was most helpful was the Nuc/Rad Communications Working Group. A tagline for the campaign was agreed upon by nuclear and media experts who participated in these conferences; "Get Inside. Stay Inside. Stay Tuned." A number of tools were intended to be created to promulgate this slogan.

Not long into the planning process I learned from Cass Kaufman, the Director of the Radiation Management Program for the Los Angeles Department of Public Health, that FEMA was interested in finding a local partner to do just what we were working on. It seemed that FEMA was trying to develop a national program for pre-nuclear explosion public information over the last couple of years but when they brought their ideas to the focus group stage, participants could not get past asking "What do you know that we don't know?" The project bogged down and then stopped altogether. Thereafter, FEMA began looking for a locality that was interested in doing this, since some backwater would be less likely to know top secret information like the threat of an impending nuclear attack and thus the public would be more easily satisfied that any such planning was based on prudence, and not a known threat. Implicit in the arrangement, I assumed though it was never explicitly stated, that once a locality broke the ice and launched a public information campaign, FEMA would do it on a national scale.

I called FEMA, and Ventura County became that backwater. Over the next year or two FEMA invested what I suspect was between $500,000 and $750,000 in Ventura County's efforts. All of this money was spent on Ventura County in the form of FEMA hired and paid-for consultants, and for three of the four videos we produced.

The first video was paid for from a separate $25,000 grant from the Centers for Disease Control.

When the first grant from FEMA was at its end, we were told it was going to be extended. We began to work with the new consultants. We sent FEMA the four videos we had made. We were told that the videos we had produced were not what FEMA was looking for. An e-mail from an important leader in FEMA raised concerns about the videos and invited me to work with FEMA's External Affairs people. He gave me the name of who I should contact. I reached out to that person but got no response. A contact told me that FEMA put the kibosh on further support because of the External Affairs people. He went on to say that FEMA operates in silos and that the External Affairs people had not been aware of the grant support that FEMA was giving Ventura County; further, that the FEMA External Affairs group felt that they had been burned by Katrina and never recovered and that they did not want to be associated with anything controversial that was elective. A must do was one thing. An optional thing was another. Finally, to diffuse the impact of vivid video messaging, FEMA External Affairs would prefer to portray the characters in their videos as stick figures. We were told that FEMA wanted its name and logo removed from all of our videos.

And so we went on for the next months without any further FEMA funding.

After many delays, we finally set a date for the launch—September 18, 2013. As I contemplated the preparations that needed to be completed prior to the launch, I found the number of moving parts intimidating. The only one besides me who knew the campaign inside and out and had the know-how and energy to put it all together and pull it off was JAMS Productions. But it would take a lot of concentrated effort to make it happen. I secured a $10,000 grant from the Health Department to pay JAMS for its last few months of work.

Since the greatest concern of our local policy makers was that people might panic on hearing about the campaign—would ask the same question "what do you know that we don't know?"—we took several steps to limit this. We built the campaign around town hall meetings. We scheduled four of these meetings in large halls spread throughout the county. The health officer gave a 15 min presentation to what was, in all four cases, about 100 members of the public and the local press. The plan was that by presenting to the public in a manner which permitted immediate response to questions and concerns raised by the audience, anxieties would be minimized. The plan worked. There was virtually no public or press outcry, fear or evident paranoia. In spite of all the concern about mass panic, all questions raised at the Town Hall Meetings were reasonable and respectful. In addition to the public, these meetings were attended by representatives from the County Board of Supervisors, the Sheriff's Department, Fire, the Air Force, State Senate, Los Angeles County Public Health and Emergency Preparedness, and Santa Barbara County Emergency Preparedness.

The campaign was announced through the press. A conscious decision was made by Public Health's administration to release news of the launch of our pre-nuclear explosion public information campaign as an exclusive to the largest Ventura County newspaper. It is my belief that our Administration also wanted to limit the awareness of this campaign to our county only. If it became a bigger story, the chances were increased that the altruism of our efforts would get confused with concerns that there might be a known nuclear threat. By giving an exclusive, Administration hoped that the campaign would get a sympathetic handling of the story which would minimize the chances of a negative public reaction. This proved to be correct. The coverage was positive and there was no significant public backlash. My concern was that by giving an exclusive to one newspaper only, other media in the county and the nation would ignore the story, there would not be the widespread saturation of the campaign information, and no one elsewhere in the nation would be encouraged to efforts of preparedness. That also proved to be correct.

One of the things that we covered in our educational efforts was the importance of helping our neighbors in the event of an IND detonation. We went to great lengths to demonstrate how to decontaminate and welcome contaminated people fleeing from an IND disaster. We taught that helping our friends and family from Los Angeles and welcoming them into our homes was a test of our humanity. And we taught how safe it was to do so.

An 18 page document, created for the general public as an educational tool, was written in both Spanish and English.

Four 1–2 min videos were created which addressed different areas of concern for the public. One spoke to parents and asked them to talk about nuclear preparedness with their children. Another focused on teaching the slogan "Get Inside. Stay Inside. Stay Tuned." A third urged parents to stay at home and trust their schools to care for their children until the concern about fallout had passed. And the fourth taught that the appropriate thing to do is not to "Duck and Cover" as used to be said in the fifties and sixties, but to "Get Inside. Stay Inside. Stay Tuned." So that the videos would not be seen out of context and frighten people, a 1 min introduction was created and added to the beginning of each. The introduction stressed that this campaign had been a part of nuclear preparedness efforts which had been going on in the county for 9 years, that there was no known threat and that this was just another program in our county's emergency preparedness portfolio. Only one video was introduced at the time of the launch. It had a humorous tone and was well received. That video PSA is the #2 most-viewed video of all time on Ventura County Public Health's YouTube page and spread faster (and shared at a rate more frequently) than any video to date.

A web site was created with a link to the educational video; separate links to extensive libraries of articles for the general public, for health professionals and for first responders; and a tab which allowed the user to request a speaker from the Public Health Department to talk to their community group. http://vchca.org/nuclear-educational-campaign

A phone bank was established and staffed by local volunteers and two subject matter experts. One was Tosh Ushino from the Health Physics Society and the other was Tom Gorman, the training coordinator from the Radiological Assistance Program. The phone bank was established on the chance that there might be questions or a significant amount of concern in the community as a result of the campaign launch. Only 20 calls were received. Most of the calls asked about where the next Town Hall Meeting would be.

A letter was sent to parents of all school children in the county warning them that a Pre-Nuclear Event Public Information Campaign was going to begin in several days, that their children would not be targeted at this point for any specific education but that the children might hear of the campaign and have questions.

A flyer was created to educate the public about the proper steps to take in the event of a nearby nuclear detonation.

The campaign garnered positive media attention from the Ventura County Star and local National Public Radio and from several national organizations and institutes. These latter included the University of Pittsburgh Medical Center, Center for Health Security (the authors of Rad Resilient City) and the United States Environmental Protection Agency. Additional local coverage came from the Moorpark Patch and the Camarillo Acorn. There were a couple of negative comments. The Huffington Post made a sarcastic comment about the letter to the parents and another tweet suggested Ventura County was jumping on the bandwagon of fear. None of these few comments developed any traction. One of the 27 Superintendents of Schools in the county was critical of the letter to the parents because the schools already had too much on their plates to be taking on nuclear preparedness. This concern was covered by the Ventura County Star but appeared to be a lone voice.

What conclusions am I left with? What remains to be done?

There is an element of nuclear planning which has haunted me from the very beginning. How do we evacuate a large population from an urban area? We ask people to get inside and stay there until they hear on their emergency station that it is okay for them to leave. But with hundreds of thousands of cars on the road there will be standstill traffic for 75–100 miles in all directions. We will be sending people out to parking lots where they will be less comfortable and more exposed to radiation than they would be at the place they just evacuated. Let's take this quandary apart to see if it is subject to remediation. How many cars does it take to stop traffic? This will occur with just a few stalled cars. We've all experienced this. We experience stalled traffic most mornings and nights on our way to and from work and usually that's just from more cars, not from disabled ones. Now imagine all the reasons why a car will stall. Running out of gas. A collision as a result of an unfortunate lane change. An old or poorly maintained car whose time has come. What are the chances that with hundreds of thousands of additional cars on the roads that this will happen? It is unthinkable that it will not. So how do we keep all these cars off the road?

We educate the public to get inside, stay inside and stay tuned to their emergency broadcast station. And with the finest education campaign ever, what percentage of the population will we reach? Seventy percent? Eighty? There's never been an education program that has penetrated that thoroughly. Not in America. Twenty-nine percent of Americans don't know who the vice president of the United States is and 6 % don't know what day Independence Day falls on. And of the seventy percent that we reach, how many will listen, will obey when the time comes? Fifty percent? They have children, parents, spouses. They're smarter than the emergency planners. They have to rescue their children. They can beat the crowds out of town. They know everyone else will listen, will obey, will stay inside, and they'll be able to get out ahead of them.

I'm going to stay put not because I'm smart but because so many other people are going to behave stupidly. How many days will those traffic jams remain? Ask yourself, how many of those cars are going to run out of gas? How many people are going to need food or a rest room and just walk away from their stopped car in the middle of the 405 freeway? Look further up the road. There will be cars that make it to Ventura but need to get off the highway for food or a rest room or gas. But the gas stations will be out of gas. And more cars will be abandoned or run out of gas on the surface streets, not only in Ventura County but in the neighboring counties, and then on the exit ramps. The log-jam of stalled cars will extend for 75 miles in every direction. And whether the California Highway Patrol wants to or not, there will be contraflow. People will leave the highway and enter the opposite direction highway on the other side of the barrier by entering on an exit ramp and making it across the highway, if they're lucky, to the highway apron. And if they're unlucky, they'll have a head-on collision and bring traffic on the other side of the highway to a standstill. Those roads will clear when bulldozers are brought out to push the vehicles off to the side of the road like so much snow.

Emergency planners call all of this congestion. I call it congealment.

Solve that problem and maybe our nuclear planning efforts will move along more smoothly.

But that doesn't mean our enemies will wait.

What other conclusions am I left with? What else must be done?

The silence has been broken. Preparing for a nuclear detonation can now be spoken of in polite company. We have shown that the matter can be discussed without causing panic. Combining this reality with the HHS Study which states that 280,000 lives can be saved where there is a prepared public, and the propensity for al Qaeda to perform multiple simultaneous attacks, we could easily be talking about saving one million lives. But every county in the nation has to have an informed public, not just Ventura County. And now throw in the evolving threat of ISIS and all of the returning US citizens who are currently fighting for them. We face a threat that is getting stronger. Who can we turn to to carry a national pre-nuclear event public information campaign forward? Who has the clout and the machinery to make this happen? The unavoidable answer is FEMA. The log-jam that they faced, public suspicion, has now been broken up. It has been shown that pre-nuclear event public information can be achieved without causing chaos and mass hysteria. It is time to act.

And it's surprising that FEMA has not acted. In a December 15, 2010 article by William J. Broad in the New York Times titled "US Rethinks Strategy for the Unthinkable", W. Craig Fugate, Administrator of FEMA, is quoted as saying about nuclear preparedness, "We have to get past the mental block that says it's too terrible to think about. We have to be ready to deal with it and help people learn how to best protect themselves."

For a moment, let's review some of the disasters that have affected the United States in the last couple of years.

16.1 Disasters Affecting the United States

2014

- Mudflow, 43 deaths, Oso Washington
- **Beheading of Westerners by ISIL**
- **Ebola virus transmitted in the United States**
- Tornado, 35 deaths, April 2014, Nebraska, Louisiana, Oklahoma, Illinois, Florida, N. Carolina
- 20 major wildfires in California; 12 located in San Diego

2013

- California wildfires, a group of 7,176 wildfires that burned at least 593,985 acres of land. The wildfires injured 125 people and killed 1.
- Tornado, 24 deaths, Moore Oklahoma, 377 people injured.
- Wildfires, Yarnell Hill, Arizona, 19 firefighters killed/trapped when they became trapped in the fast-spreading fire.
- Accident/Explosion, 15 deaths, West Fertilizer Company explosion, West, Texas, More than 160 people injured.
- **Boston Marathon bombings, 3 fatalities and an additional 264 people injured by the two separate bombings which were 13 s apart. 17 others injured in subsequent gunfight (16 police officers and a perpetrator).**
- 2013–2014, Cold wave, 21 deaths, Eastern US
- Tornado, 8 deaths, Canadian County, Oklahoma, four storm chasers killed, including Tim Samaras. First tornado on record to kill storm chasers, 151 injured, wildest tornado on record with a path width of 2.6 miles.

2012

- **Hurricane Sandy, 147 deaths, $75 billion.**
- **Mass murder, school shooting, Sandy Hook Elementary School, Connecticut, Adam Lanza shot and killed his mother, then drove to Sandy Hook, where he fatally shot 20 students and 6 educators before committing suicide.**
- Mass Murder, 12 deaths, Aurora Colorado, 56 others were injured.
- Mass murder, 7 deaths, Oikos University, Oakland, California.

All of these incidents, big and small, got more national press than nuclear preparedness. The least of them had more publicly acknowledged funds spent on them for both investigation and prevention. What do all of these events have in common? They all happened.

Does a nuclear device detonated in the United States by terrorists need to happen for us to take the relatively inexpensive and easy step of educating our public?

George Bush said of a nuclear detonation in late 2002, "History will judge harshly those who saw this coming danger but failed to act."

People say that a nuclear detonation by terrorists on American soil is a low probability, high consequence act. I say that it is inevitable with increasing probability.

This doesn't need some huge allotment from Congress. I could do this and so could 50,000 other people in this nation. The work is all done. Use the materials we created in Ventura County. They're FEMA's or anyone else's who wants to use them. Any one of us could do this with a tiny sliver of FEMA's staff and budget.

FEMA can have the most profound impact on nuclear preparedness in the United States but they refuse to act. But that doesn't mean that there is nothing we can do. Our elected officials can hand FEMA their marching orders along with a time table. Is there an elected official who wants to be our champion? A major health and medicine editor like Sanjay Gupta or Richard Besser could shame our decision makers into doing this. In the meantime, localities can prepare their own populations for a nuclear disaster and should start this process as soon as possible.

I've tried to make a coherent story of the events that led up to and through Ventura County's Pre-Nuclear Event Public Information Campaign. I have asserted that an agency with the title of the Federal Emergency Management Agency should step up and launch an effective and colorful campaign to educate the public as to what steps to take to protect itself in the event of a terrorist's nuclear detonation.

Good public health challenges old ways of doing things. The best public health is usually outright controversial. Leaders who are either elected or are appointed and serve at someone's pleasure are risk-averse by nature. The majority of the electorate respects public health officials and their initiatives. So what are our officials so afraid of?

Now I bring this battle to you. You all work on nuclear preparedness. Each of us is a Sisyphus in his or her own way. The years go by and, it seems, little progress is made. It's exhausting. But if you're tired of fighting and you're right, who loses?

Chapter 17
The Romanian Management System for Nuclear and Radiological Emergencies

Mihaela Mihaila

Abstract At national level, **The General Inspectorate for Emergency Situations (GIES)** is the specialized body of the Romanian Ministry of Internal Affairs responsible for the coordination of all organizations involved in emergency situations management, as part of the National Emergency Management System (NEMS) and component of the National Defence System. The National Emergency Management System aims at prevention and management of emergency situations, the planning and coordination of human, material and financial resources. The NEMS is an integrated framework inside of which all the support tasks for prevention and management of emergencies are shared among national ministries, central bodies and non-governmental organizations.

The response is developed based on the National Plan for Protection and Intervention in case of a nuclear accident or radiological emergency, providing a national, coordinated response of all the components of the National Emergency System.

GENERAL INFORMATION ABOUT NPPs affecting Romania

Romania has one nuclear power plant, Cernavoda NPP, with two units in operation, pressurised heavy water reactors of CANDU 6 design (CANadian Deuterium Uranium), each with a design gross output of 706.5 MWe.

In Romania there is also a nuclear research reactor (Tryga) and, on the Romanian Bulgarian border, the Bulgarian State has the Kozloduy nuclear power plant. Kozloduy NPP currently manages two pressurized water reactors with a total output of 2,000 MWe.

In case of an accident, the response starts from the lowest level raising to the national level, if needed. Under the GIES, are established the county inspectorates (42) and in the last 5 years, they were endowed with new techniques, equipment and special training of first responders which will be used in the field response.

The spread of these capabilities around the country takes into account thefollowing criteria:

M. Mihaila (✉)
General Inspectorate for Emergency Situations, Ministry of Internal Affairs, Bucharest, Romania
e-mail: duditam@mai.gov.ro

S. Apikyan, D. Diamond (eds.), *Nuclear Terrorism and National Preparedness*,
NATO Science for Peace and Security Series B: Physics and Biophysics,
DOI 10.1007/978-94-017-9891-4_17

- The territorial risks;
- The financial capacity of the county to maintain and improve the technical equipment;
- The performances, endowments and the autonomy of the equipment;
- The existence of prepared personnel for these CBRN equipment;
- The readiness for international assistance missions according to the EU, NATO and bilateral agreements.

Taking into account all these criteria, the optimal repartition of the equipment around the country was chosen, thus covering, from technical point of view, the entire response necessity to a nuclear or radiological emergency.

17.1 Introduction

Emergency preparedness and response in Romania is organised in accordance with the Law 15/2005 for the approval of the Governmental Ordinance no. 21/2004, regarding the National Emergency Management System (SNMSU) and Law 111/1996 regarding to the safe deployment, regulation, licensing and control of nuclear activities.

Specific Regulations are in place in the field of radiation emergency preparedness and response:

- Nuclear Safety Requirements on Emergency Plans, Preparedness and Intervention for Nuclear Accidents and Radiological Emergencies (approved by Ministerial Order No. 242/1993);
- Fundamental Requirements on Radiological Safety (NFSR);
- Governmental Decree 223/1990 for the Romania's accession to the IAEA's Conventions on Early Notification of a Nuclear Accident and on Assistance in the Case of a Nuclear Accident or Radiological Emergency;
- Bilateral early notification agreements with Bulgaria, Greece, Hungary, Slovakia, Russian Federation, Turkey and Ukraine.

In June 2009 European Commission adopted the EU CBRN Action Plan which intends to focus on three main areas of action:

- ensuring that unauthorized access to CBRN materials of concern is as difficult as possible – **prevention**;
- having the capability to detect CBRN materials if control over them is lost – **detection**;
- being able to efficiently respond to incidents involving CBRN materials and recover from them as quickly as possible – **preparedness and response**.

Under this plan of action, Romania implemented many improvements of the response capabilities in order to answer to any nuclear and radiological emergency affecting the country.

17.2 The National Emergency Management System – NEMS

According to the current legislation, the National Emergency Management System (SNMSU) is composed from three types of structures:

- the decisional structure – the committees for emergency situations,
- the executive structure – the **General Inspectorate for Emergency Situations** (IGSU) and the county and local inspectorates for emergency situations (as public professional emergency services), and
- the operational structure – the operative centres for emergencies.

All the decisional, executive and operational structures are established on three levels: national, county and local.

The National Emergency Management System is a nationally owned mechanism of multi-stakeholders which provides coordination and response in case of emergencies, and serves as an advocate for prevention and disaster risk reduction at different levels.

- **Basic principles** of the system are:
  - Proactiveness & prevention;
  - Priority of human life protection;
  - Local authority responsibility for disaster management;
  - Gradual reaction.
- **Basic elements** of the Romanian emergency system:
  - Committees for emergency situations;
  - **The General Inspectorate for Emergency Situations – IGSU;**
  - Professional (County inspectorates) & volunteer intervention structures;
  - Emergency Operation Centers;
  - Incident Commander.

As a decision structure, at national level is organized the National Committee for Emergency Situations (CNSU). The National Committee for Emergency Situations is set-up under the co-ordination of the Prime Minister and managed by the Minister of Internal Affairs. All the ministerial, county and local committees are subordinated to the National Committee for Emergency Situations (Fig. 17.1).

17.3 The General Inspectorate for Emergency Situations – IGSU

As an executive structure, at national level is established the General Inspectorate for Emergency Situations (IGSU), a specialized organization in the Ministry of Internal Affairs. It handles in a unitary conception theimplementation of specific

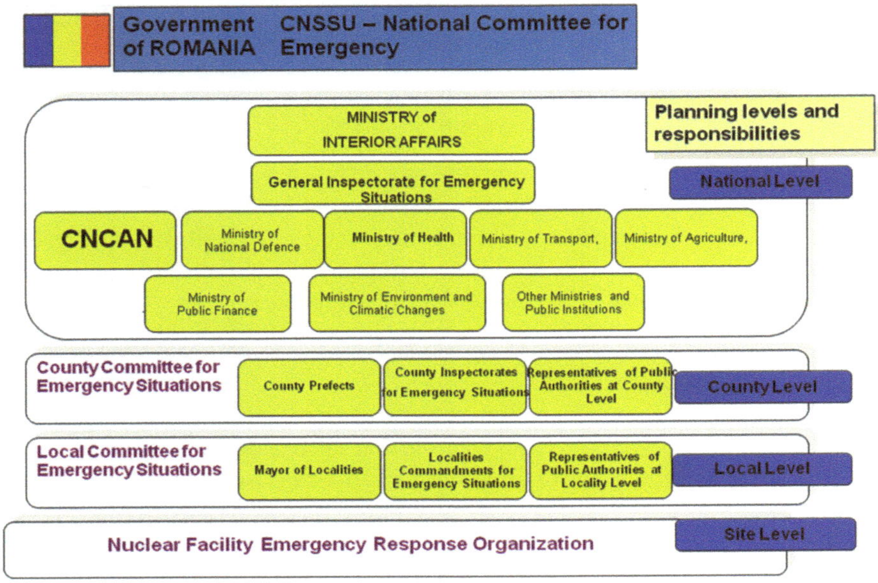

Fig. 17.1 National Emergency Management System (NEMS)

legislation, at national level, in the field of civil emergencies, protection of life, goods and environment against fires and disasters, as well as the accomplishment of civil protection measures and management of emergency situations.

By law, IGSU is the national POC for the relevant authorities (governmental and non-governmental) and international organizations acting in the field of civil emergencies. It provides and develops the international cooperation, as well as the management of international assistance requested or received by Romania.

The main tasks of IGSU are:

- to coordinate the unitary implementation of emergency management actions and measures on the national territory.
- to coordinate the response actions performed by the professional services.
- to disseminate the decisions of the National Committee for Emergency Situations.
- to provide the public information, through media, about the emergency situation status and the measures undertaken in order to mitigate the consequences.
- to collaborate and cooperate with the other state bodies with responsibilities in the management of emergency situations, state of siege or other exceptional situations.
- to manage a 24/7 Point of Contact that handles all national civil emergencies and international disasters.

At the county/local level there are established County/Local Committees for Emergency Situations, which are directed by the county Prefect/ Local Mayor.

Based on the legislative framework, at local level the intervention is coordinated by the Local/County Committees for Emergency Situations and performed by the Local/County Inspectorates for Emergency Situations. When the emergency situation cannot be solved by the local/county authorities, the national level (CNSU) is activated, in order to support the local intervention. Written agreements and protocols are in place between the responsible organizations, at local and national level, for common activities and exchange of information in emergency situations.

17.4 Overview of Nuclear/Radiological Facilities and Activities

The inventory of radiation sources and practices has not been fully completed, but it is in progress. The threat categorization will follow the international requirements.

An on site emergency intervention Plan of each nuclear installation establishes a classification system for abnormal events which specifies response actions on the basis of the actual or potential consequences of an incident for public, environment, personnel and property.

For threat categories IV and V, threat assessment is part of a national assessment. Considering the categories of radiation-related threats for the purpose of emergency preparedness and response, Romania has facilities and activities in all threat categories. A thorough threat assessment to address all types of nuclear and radiological threats in Romania was not presented (Fig. 17.2).

Threat category	Facility
I	Cernavoda NPP with two CANDU reactors
II	TRIGA reactor in Pitesti (14 MWth)
III	Facilities on the site of the Institute for Nuclear Research (INR) in Pitesti, e.g. the nuclear fuel factory, the gamma irradiation facility and the radioactive waste treatment section
IV	Different types of activities throughout the country with the use radioactive sources (e.g. medical, industrial radiography, etc.)
V	Applicable to Romania also

Romania has a only one nuclear power plant, Cernavoda NPP, with two units in operation. The Cernavoda NPP with two Canada Deuterium Uranium (CANDU) reactors is a Threat Category I facility. Work on the Cernavoda NPP site started in 1980 with Unit 1 and in 1982 with the other 4 units. Unit 1 was completed in 1996 and Unit 2 was put into operation in 2007. The Government has plans to further increase nuclear generating capacity through the commissioning of Units 3 and 4 of the Cernavoda NPP. The decision to complete Units 3 and 4 was taken in June 2007. Pre-licensing reviews have been successfully completed, but no application for a construction licence has been submitted yet.

Fig. 17.2 Authorized nuclear and radiological installation from Romania

The Kozloduy NPP is located in the vicinity of the Romanian border and its emergency planning zones spread out also onto Romanian territory. It is expected that threat assessments for the Romanian territory for possible consequences due to accidents in the water-cooled, water moderated energy reactors (WWER) type of reactors will be done in the vicinity of the Kozloduy NPP.

Romania's nuclear fuel cycle comprises a TRIGA research reactor in Pitesti, the Cernavoda NPP, the nuclear fuel plant and the national repository for low and intermediate level radioactive waste.

The 14 MW Material Testing and Research Reactor (TRIGA) in Pitesti is a Threat Category II facility. There are also other facilities on the site of the Institute for Nuclear Research (INR) in Pitesti, e.g. the nuclear fuel factory, the gamma irradiation facility and the radioactive waste treatment section, which are classified as Threat Category III facilities.

Moreover, there are different types of activities throughout the country with the use of radioactive sources (e.g. medical, industrial radiography, etc.) which belong to Threat Categories III or IV. Threat Category V is also applicable to Romania.

17.5 Overview of Preparedness Elements

The Nuclear Law No.111/1996 republished, on the safe deployment, regulation, authorization and control of nuclear activities stipulates that National Commission for Nuclear Activities Control (CNCAN) is the national competent authority in the nuclear field, with duties in regulation, authorization and control of all nuclear activities in Romania.

The intervention plans in case of radiological emergencies, caused by nuclear accidents in NPPs located on the territory of other states that may affect the Romanian territory, by transboundary effects, as well as the general off-site intervention plans for nuclear plants on the Romanian territory are prepared by the General Inspectorate for Emergency Situations (IGSU). These general intervention plans are submitted for approval to CNSSU and their applicability has to be periodically assessed and controlled by IGSU. The central and local public authorities tasked in the field of preparedness and practical response to a nuclear accident are responsible for developing their own plans correlated with the general intervention plan. These plans must be approved by the respective authorities, with the advice of IGSU, and their applicability has to be periodically assessed and controlled by IGSU.

Nuclear Safety Requirements on Emergency Plans, Preparedness and Intervention for Nuclear Accidents and Radiological Emergencies, approved by Ministerial Order No. 242/1993 (this regulation will be further referred to as MO 242/1993) are establishing the specific actions to be taken by the operator, competent authorities and other responsible public authorities for planning, preparedness and intervention in the following cases:

- nuclear accidents at nuclear installations;
- radiological emergencies resulted from licensed practices;
- radiological emergencies resulted from transboundary effects.

According to these requirements, any operator of a nuclear installation has to make preparations, in conjunction with national, regional and local public authorities and support organisations, for coping with nuclear accidents. Also, a General Emergency Plan has to be updated for any nuclear risk area in the country, which may be threatened by a radiation emergency. This Plan cover all activities planned to be carried out by all responsible authorities and organisations involved in case of an emergency situation leading to, or likely to lead to, a significant release of radioactivity beyond the site boundary of the nuclear facility.

The last version of the National Emergency Plan is in place since 2009, according to the legislative framework. This Plan cover all activities planned to be carried out by all authorities and organizations involved in case of an emergency situation leading to, or likely to lead to, a significant release of radioactivity beyond the boundary of the nuclear facility. This overall plan includes provisions for the co-ordination of emergency plans of the operating organization and of the public authorities. The National Emergency Plan is elaborated by IGSU.

As per law, each public authority has to prepare his own Emergency Intervention Plan that has to be approved by his leader, and shall be endorsed by the General Inspectorate for Emergency Situations (IGSU). The applicability of the public authority's plans is periodically controlled and evaluated by IGSU.

IGSU and National Commission for Nuclear Activities Control (CNCAN) are, according to the legislation, the national competent authorities in case of nuclear accident or radiological emergency.

Important training courses and exercises (national and international) were conducted in the last years in the field of radiation emergency preparedness and response.

The training of the staff is tested in the preparation and participation to yearly exercises organized by the major nuclear objectives in Romania.

Once in a few years, all the responsible organizations participate in the national large scale exercise organized by IGSU. The frequency of the training and exercises became constant in the last 3–4 years with at least one major international exercise and one major national exercise being organized by CNCAN in partnership with national and international institutions.

Internal (inside each organization) training courses and exercises are performed by the nuclear facilities, based on pre-established plans and programs. The licensee shall ensure the adequate initial and periodical training for the personnel authorized to declare emergency situations and to manage the intervention, for the personnel responsible for assessing the emergency, for the personnel of teams assigned for radiological monitoring and decontamination, for the personnel of control room and for field operators, fire fighting teams, repair teams and those assignees for evaluation of damages, rescue and first-aid teams. The personnel assigned for emergency response shall be regularly trained, at least every 3 months. The licensee has to maintain and verify the training of its own personnel by organizing annual exercises. The exercises shall be planned such that they cover all the seasons and all meteorological conditions. All the exercises shall be followed by an evaluation process, with the participation of representatives of the competent authorities.

The exercises end with an analysis and a balance of activities in order to evaluate the ability of the various components/organizations involved. The deficiencies noted during the exercises are to be corrected by appropriate training and/or other corrective actions.

Provisions are included in the Emergency Response Plans for ensuring the availability and reliability of all supplies, equipment, communication systems and facilities needed during an emergency.

IGSU is asking all the partner institutions to provide annual plans with necessary material and human resources for emergency preparedness and response actions, and also annual plans with the activities of the Ministerial Committees of the responsible organizations.

The financing of the competent authorities is assured for each responsible institution by the annual approved state budget, as a distinctive part of the annual budget of each state institution.

17.6 Overview of Response Elements Establishing Emergency Management and Operations

By law, the Ministry of Internal Affairs (MAI) through IGSU is responsible for the management of nuclear and radiological emergencies. As national competent authority, CNCAN is acting as the technical adviser of national decision makers for nuclear safety analysis and for radiation protection issues, in case of a radiation emergency.

According to the current legislative framework, at local level the intervention is coordinated by the Local/County Committees for Emergencies and performed by the Local/County Inspectorates for Emergencies. When the emergency situation cannot be solved by the local authorities, the national level is activated (CNSU through IGSU), in order to support the local intervention.

Written agreements and protocols are in place between the responsible organizations, at local and central level, for common activities and exchange of information in emergency situations.

17.6.1 Identifying, Notifying and Activating

At national level, there are established two National Contact Points: one in relation to European Union – ECURIE system and one in relation with IAEA – ENAC system for early notification:

1. The National Contact Point in relation to ECURIE system is organized under the General Inspectorate for Emergency Situations.
2. CNCAN is the National Contact Point in relation with IAEA, in respect with the provisions of IAEA Conventions for Early Notification and Assistance (Law 111/1996 amended in 2003 and IAEA letter EPR/CP(0100) from 16/11/2000).

Inside each Inspectorate for Emergency Situations is set-up an Emergency Operation Centre (EOC) for, with permanent activity, ready to activate the emergency organization in case of an accidental event. These EOCs receive notifications for all types of emergency, including radiation events. Also, the responsible organizations at national level operate such EOCs, in accordance with the legal provisions in their field of activity.

The notification system is established in the Emergency Plans of the authorization holders and public authorities. Exercises and communication tests are performed between the operative centres of the National System.

In case of a nuclear incident/accident at a nuclear facility, the response actions should begin without any delay and be coordinated from the start. To facilitate this, an event classification system is established by the On Site Emergency Intervention Plan of each nuclear installation, in order to predefine the response actions for each emergency class. The events are classified on the basis of the actual or potential consequences of an incident for the public, environment, personnel and property. In case of radiation emergencies with off-site effects, the operator is responsible for initiating notification of the public authorities and for elaborating first recommendations on protective actions for the population in the affected area.

17.6.2 Taking Mitigatory Actions

The County Emergency Plans for Radiological Accidents specify the way to obtain expertise and services in radiation protection field, at local level, in a timely manner.

Provisions are in place for Cernavoda NPP and these provisions are included in the On-site Emergency Plan of Cernavoda NPP and also in the National Intervention Plan for Cernavoda NPP.

In case of accidents with releases of radioactivity material in the environment, the operator is also responsible to determine the amount of radioactivity released.

Provisions are in place for Kozlodui NPP and these provisions are included in the General Intervention Plan for Dolj – Bechet area (the most exposed Romanian area in the vicinity of Kozlodui NPP). The provisions in Plan are related to mitigation of the consequences of an accidental release or exposure from Kozlodui NPP.

17.6.3 Taking Urgent Protective Actions

In order to take protective actions for the affected population, there is established, at national level, the National Centre for Intervention Co-ordination in case of a nuclear accident or radiological emergency (CNCI), where representatives of central public authorities are activated in case of an emergency. In this National Centre is functioning, in case of a severe nuclear accident, a central commission of experts (CANUR) that has the responsibility of assessing the emergency and give technical recommendations to national decision makers.

The automatic connection of CNCI with the operative centres of the responsible organizations is to be established, for data transfer and exchange of information between local and central levels.

Generic intervention levels are defined in the current:

Protective measure	Dose quantity	Dose criteria
Sheltering	Committed effective dose	10 mSv for the first 7 days
Evacuation	Committed effective dose	100 mSv for the first 7 days
Iodine	Committed effective dose	50 mSv for the first 7 days
Relocation	Committed effective dose	30 mSv for the following 30 days after the first 7

17.6.4 Providing Information and Issuing Instructions

Arrangements are in place in all nuclear risk areas (Cernavoda NPP, Bechet area – near Kozlodui NPP, TRIGA reactor in Pitesti – Mioveni) for prompt warning and instruction of population in the emergency planning zone, in case of an accidental event.

In the first phase of a severe accident, the operator of the nuclear facility is responsible for giving first instructions to the population in the affected area. The protective actions to be recommended for the population by the operator are established based on the projected doses and the off-site ambient dose rates. In normal conditions, the public in the vicinity of Cernavoda NPP and Kozlodui NPP receive periodically printed information about the threat and how to behave in the case of an emergency. At local/county level, a Public Information Group is established in case of emergency in order to provide information to mass-media and population.

17.6.5 Protecting Emergency Workers

Provisions are in place in the current legislation for radiological protection of personnel. New provisions (generic action levels for the intervention personnel) will be adopted when MO 242/1993 will be revised, accordingly with the international recommendations.

In the Fundamental Requirements on Radiological Safety (NFSR), issued by CNCAN 2000, are stipulated the dose limits for workers:

- 20 mSv per year for whole body,
- 150 mSv per year for the lens of the eye (dose equivalent),
- 500 mSv per year to the skin of the whole body or to the skin of any extremity (dose equivalent).

For young people under 16, working with radiation, we have established a dose limit as for population, of 1 mSv per year. For people between 16 and 18 years, we have established a dose limit of 6 mSv per year.

There are provisions related to the dose limits for workers in intervention:

- Hp(10) <100 mSv – in order to prevent large collective dose
- Hp(10) <250 mSv – in order to prevent deterministic
- Hp(10) <250 mSv – in special situations for life saving actions and in actions performed to prevent or stop the development of catastrophic conditions in installation.

17.6.6 Assessing the Initial Phase

No OILs are established yet at national level.

Nevertheless, according to the OM 242/1993 provisions, under the emergency intervention plans, optimized intervention levels shall be established, based on particular locations and circumstances of each nuclear installation, taking into consideration radiological factors, as well as financial and socio-economical factors.

Operational Intervention Levels are established for Cernavoda NPP under the On Site Intervention Plans, in terms of measured values of measurable quantities (dose rates, activity concentrations, etc.). The model and hypotheses used to derive these Operational, Intervention Levels (OIL) are described under the On Site Intervention Plan.

In the first phase of the emergency, the protective actions are established by comparing the measured dose rates with the OIL's calculated during the emergency planning process. After the information about the emergency conditions and about the source term released into environment becomes available, some OILs are recalculated, according to specific health physics procedures. Then, protective actions are established comparing the measured dose rates with the new values of the OILs.

The following OILs are established for Cernavoda NPP, considering as generic intervention levels the averted doses recommended by IAEA.

OIL	Value	Protective action
OIL 1	1 mSv/h[(a),(c)]	Evacuate or provide substantial shelter[(b)] for this sector, the adjacent sectors and the sectors closer to the plant. Until evacuated, people should be instructed to stay inside, with their windows closed.
OIL 2	0.1 mSv/h[(c)]	Take thyroid blocking agent, go inside, close windows and doors and monitor radio and TV for further instructions.

[(a)] If there is no indication of core damage, OIL 1 = 10 mSv/h

[(b)] "Substantial shelter" is provided by specially designed shelters or the inside halls or basements of large masonry buildings. Shelter should be considered only for 24–48 h and its effectiveness must be confirmed by monitoring, especially in high dose rate areas

[(c)] Monitor evacuees and instruct the public on decontamination measures

OIL	Value	Protective action
OIL 3	1 mSv/h	Evacuate or provide substantial shelter within the sector
OIL 4	0.2 mSv/h[(a),(b)]	Consider relocating people from the sector
OIL 5	0.001 mSv/h	Restrict immediate consumption of potentially contaminated food and milk in the area, until samples are evaluated

[(a)] This OIL has to be recalculate based on sample analysis as soon as possible

[(b)] For 2–7 days after the accident

17.6.7 Managing the Medical Response

Arrangements have been made in the last years for general practitioners and emergency staff to be made aware of the medical symptoms of radiation exposure and of the appropriate notification procedures if a nuclear or radiological emergency is suspected, and there are irregular training courses.

The Polyclinic of Cernavoda (near the NPP), and County Hospital in Constanta have been prepared to treat injured people, as a consequence of a radiation event at Cernavoda NPP.

At national level, there is established a place for initial treatment of overexposed people at the Clinique for Radiopathology belonging to the Institute for Public Health in Bucharest.

17.6.8 Keeping the Public Informed

The On-site Radiation Emergency Plan of the operator and the Off-site Radiation Emergency Plans of the public authorities establish the responsibilities, the resources and the interfaces required for informing the public in case of a nuclear emergency. Joint information centres, staffed by representatives of the nuclear facility and of the public authorities, are established at the local and national levels.

At local level, the information includes instructions and warnings for the population in the affected area.

CNSSU, at national level, and the County Committees for Emergencies, at local level, are responsible to give instructions and information to the public. The local and national TV and mass-media are used to keep the public informed about the accidental radiological event.

At national level, the information includes aspects regarding the status of the nuclear/radiological facility and to the status of planning/implementing the protective actions for population.

17.6.9 Taking Agricultural Countermeasures Against Ingestion and Longer Term Protective Actions

Current provisions are stipulated in MO 242/1993 for taking long-term protective actions. Generic action levels for foodstuffs are established. Provisions are stipulated in the National Intervention Plan for Cernavoda NPP and also in the County Intervention Plan for Dolj – Bechet area in relation to the implementation of long-term protective actions.

17.6.10 Mitigating the Non-radiological Consequences of the Emergency and the Response

CNSSU, at national level, and County Committees for Emergencies, at local county and level, are responsible to give instructions and information to the public. The local and national TV and mass-media are used to keep the public informed about the accidental event.

Educational programs are in place in the nuclear risks areas in order to explain to the population what the risks are and what has to do in case of a radiation emergency, this contributing to the mitigation of any non-radiological consequence during an accidental event.

17.7 Challenges in EPR

Some organizations were, in the last years, the beneficiary of European funds, but the National System for the Management of Emergencies still need big financial efforts in human resources, documentation, equipment, communication systems, facilities, supplies. Ongoing major national activities are related to development of infrastructure for data transfer and exchange of information between the operative centres of the National System. The National Emergency Response Plan is under revision.

There is a need for future developments and improvements at all levels within the national system for the management of emergency situations.

There is a need for developing and harmonizing at national level the technical capabilities for assessing the radiological consequences of any radiation event.

There is also a need to develop and harmonize the communication arrangements in between the emergency response centres of all organizations belonging to the national system for the management of emergency situations, now being in place a not very efficient communication system between all the authorities involved in the response because of:

- Lack of common language,
- Equipment compatibility,
- Not clear identification of the essential information for different actors involved in the response.

Another challenging area is the **Public information**. It is known that raising the education and awareness can contribute to minimising the effects of nuclear or radiological emergencies on the citizens and to help the citizens to protect themselves more effectively. An important aspect of increasing the security is an educated and well informed public. Consequently, there is a need to undertake further work on improving communication with the public. Many States have already developed specific communication strategies and can be very useful to exchange experience with their experts.

Part IV
Radiation, Associated Risks and Public Health

Chapter 18
Initiative to Enhance Radiological Resilience

Kai Vetter

Abstract The recent events at the Dai-ichi nuclear power plant in Fukushima, Japan, have highlighted the need to enhance the resilience to radiological and nuclear events. The incidence and the associated large releases of radioactive materials had and still have an enormous societal and economical impact in Japan and globally. Although no casualties and health effects have been and likely will be attributable directly to radiation, these events have manifold and substantial impact on local communities and have provoked ongoing anxiety and concerns globally.

We have established a new initiative on radiological resilience in Berkeley in collaboration with Japanese partners to address the needs for better scientific and technological capabilities to assess, predict, and minimize the impact of radiological contamination and to enhance the understanding of radiation and associated risks in the public. We will motivate the need for such an initiative and the associated opportunities and challenges in Japan and globally.

18.1 The Challenge

The events at the Dai-ichi nuclear power plant in Fukushima, Japan, have highlighted the need to enhance the resilience to radiological and nuclear events. The incidence and the associated large releases of radioactive materials had and still have an enormous societal and economical impact in Japan and globally. Although no casualties and health effects have been and likely will be attributable directly to radiation, these events have manifold and substantial impact on local communities and have provoked ongoing anxiety and concerns globally [1].

In order to minimize the impact of these and possible future radiological incidents, technologies and scientific understanding have to be enhanced and equally important, the understanding of nuclear and radiological matters in the public. An

K. Vetter (✉)
Nuclear Science Division, Lawrence Berkeley National Laboratory, Berkeley, CA, USA

Department of Nuclear Engineering, University of California, Berkeley, Berkeley, CA, USA
e-mail: kvetter@berkeley.edu

S. Apikyan, D. Diamond (eds.), *Nuclear Terrorism and National Preparedness*,
NATO Science for Peace and Security Series B: Physics and Biophysics,
DOI 10.1007/978-94-017-9891-4_18

important reason for concerns and anxiety in local and global communities can be found in the lack of knowledge and accessible information about radiation and the lack of clear and transparent communications resulting in an evolving distrust of the authorities and potentially of the scientific community. In addition, more effective technologies are required in combination with better scientific models to assess and predict the distribution and transport of radioisotopes in the environment and ultimately, to understand the transport into and minimize the impact onto the biosphere.

The events in Fukushima underscore the necessity and provide the opportunity to address both aspects, the need for better science and technologies and for better understanding in and communication with the public. Currently, local communities and global societies are vulnerable to such events and exposed to the actual and physical as much as to the perceived risks associated with nuclear and radiological incidents. Major releases of radioactivity will have a global impact for two reasons: (1) The radioactive materials can be transported globally quite effectively as has been observed after Chernobyl or Fukushima or previously with the nuclear weapons program; (2) Any event related to the releases of radioactivity will end up as headlines in the global and social media as it is seen news-worthy and will be – in most cases wrongly – associated with a significant health impact. Both aspects will cause increased concerns and fear world-wide, which can only be addressed by an enhanced understanding of radiation and the associated risk for environmental or biological and health effects. Furthermore, the perceived risk and perceived danger of radiation is and will hamper public acceptance of nuclear power which can contribute to a mix of carbon free energy sources necessary to meet the increasing future energy demands while reducing the global CO_2 footprint.

The concept of resilience, i.e. the ability to recover quickly from a disruption, appears provocative to the public, as it implies the possibility for more disruptions or accidents to happen. However, this is exactly what has to be realized, the fact that there is a finite probability for more accidents to happen. We need to enhance our resilience in order to be better prepared so we can minimize the physical and "measurable" impact as well as the psychological and emotional health impact. Only an informed public and educated decision-makers are able to provide an effective response to a disruption. The necessity for a better-informed public on radiological matter or more broadly for a "technologically literate citizen" is not new [2]. As we can see from the events associated with Fukushima or more recently associated with the Ebola outbreak in West Africa, the U.S. and global society are susceptible to a perceived risk rather that a factual risk. The Ebola outbreak had an enormous impact in the media and received significant attention by decision makers in the U.S. While the impact of Ebola in the originating countries was and still is enormous, in the U.S. less than 5 person got infected and the only person who died was infected in Africa [3]. Similar to Fukushima, both events resulted in a significant physical and socio-economical impact locally and mainly a – not-to-be underestimated – psychological effect elsewhere. Not to diminish the local effects, but, just to put the impact in the U.S. in context: About 150 people died of influenza and pneumonia every day in 2012 and about 15 per day in California alone [4].

18.2 The Approach

In order to enhance the resilience for current and future radiological incidents, we have established a new activity that consists of a science and technology component and an outreach and education component.

18.2.1 Outreach – The Berkeley Radwatch Project

With the releases of radioactive materials on and shortly after March 15, 2011 from the Dai-ichi nuclear power plant we set up initially rudimental, later more sophisticated means to collect rain water and air samples on the top Etcheverry Hall on UC Berkeley campus. The goal of this activity was twofold: (1) To study the characteristics such as type and quantity as well as the appearance and the disappearance of radioactive materials that we could associate with the releases in Fukushima; (2) To publish our results and to engage the public in a dialogue about radiation and to put our findings in the context of the radiation we are exposed to naturally or electively on a daily basis (Fig. 18.1).

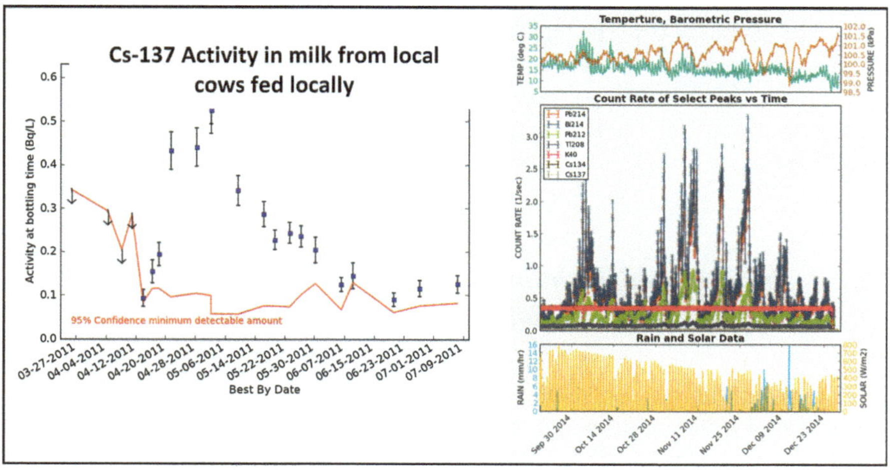

Fig. 18.1 Example of measurements and data displayed at http://radwatch.berkeley.edu. *Left*: The appearance and disappearance of Cs-137 in milk collected locally. The measurements were performed with a high-purity germanium gamma-ray spectrometer. It is important to note that the activity due to naturally occurring K-40 is about 50 Bq/l, e.g. about 10 times higher than the highest Cs activity measured once. The red line reflects the detection limit. *Right*: Near-real time display of radiological and meteorological data collected on UC Berkeley campus. The radioisotopes are observed with a particulate filter in front of a high-purity germanium detector, the meteorological data are collected with a weather station located close to the gamma-ray spectrometer. Of importance are the observations that we are continuously exposed to decay products of uranium and thorium as well as potassium (specifically K-40) and that the activity levels can vary by a factor of 10

Our ongoing measurements at UC Berkeley are part of the Radwatch project. We have established a website where we publish and discuss our results as well as claims and results from elsewhere [4]. Our measurements include a large range of environmental and food samples. Early in 2014, we installed an automatic and near-realtime air monitor that provides activities of radioactive particulates captured in a filter mounted in front of a high-energy resolution high-purity gamma-ray spectrometer on an hourly basis. Associated with our Radwatch activity, we have established the Kelpwatch project in collaboration with Steve Manley from California State University in Long Beach, CA [5]. The goal of this activity is the measurement of radioisotopes in marine Kelp that is collected along the Pacific Coast of North America. Similar to the arrival of radioisotopes from Japan that arrived in California by means of the jet stream within about 70 h after the releases into the atmosphere in Japan, the main goal of Kelpwatch is to observe the arrival of radioactive materials that were released into the Ocean in March 2011 by means of the Pacific Ocean currents. While we expect the arrival of radioisotopes such as Cs-134 and Cs-137 that can be associated with the releases in Japan over the next year or so, the concentrations will be extremely small, close to the detection limit of our monitoring system of about 10 mBq/l and will not pose any health risk [6–8]. The Woods Whole Oceanographic Institution recently published the first observation of Cs-134 off the Northern California Coast. The observed concentration of 2 mBq/l is significantly less than the about 10 Bq/l of the naturally occurring K-40 observed in the Pacific Ocean [9]. The ongoing releases of contaminated water in the Ocean in Japan result only in fairly small concentrations even close to the harbor of the power plant. For example, the water concentration of Cs-134 in close proximity to the harbor of the Dai-ichi nuclear power plant is about 20 mBq/l, if detectable at all [10].

18.2.2 Science and Technology – Assess, Predict, and Minimize the Impact of Radiological Contamination

Our research within the context of radiological resilience at LBNL is currently engaged in 4 different scientific and technology domains, which address current needs in the evacuated areas to ensure the safety of the population when they return and in the future. As of October 2014, about 85,000 people are still evacuated and many, particularly older people want to move back to their homes and communities [11]. The research areas we are currently pursuing in collaboration with scientists from JAEA aim at more effective means to map the contamination, at a better understanding and improved predictive power of environmental transport models, enhanced understanding and measurements of internal human radiation dose, and at the removal of contamination from soil. While the focus initially is on the most abundant contaminant left, cesium, and the environment of Fukushima, the knowledge gained and technologies developed will provide significantly improved

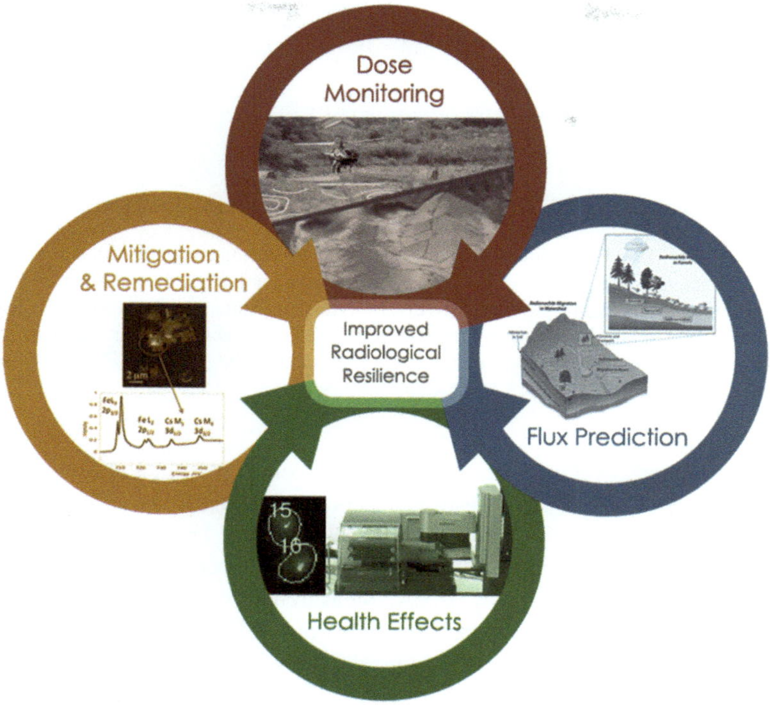

Fig. 18.2 The four main research areas currently pursued as part of the Berkeley Initiative for Radiological Resilience. The goal is to enhance the effectiveness in monitoring and predicting radiological transport in the environment, to better understand and minimize the impact of radioisotopes in the biospheres, particularly humans, and the remediation of these radiological materials, particularly cesium

means in the aftermath of any radiological event in the future that is associated with the release of radioactive materials. The Fukushima Prefecture represents a very different environment than for example the region of Chernobyl as it consists of large portions of forests and mountains with significant precipitation over the whole year causing continuous changes in the contamination patterns. Figure 18.2 summarizes the four areas of research and their relationships. These activities are coordinated with the substantial efforts by the JAEA in Japan.

18.2.2.1 Effective Contamination Mapping and Monitoring

The effective and sustained mapping of radioisotope contamination in complex, three-dimensional topographies is critical for predictive modeling and guidance of response and remediation activities. We continue to develop new detection concepts and systems that combine sensitive radiation detection and imaging instruments with complementary environmental sensors. These systems are based on a variety

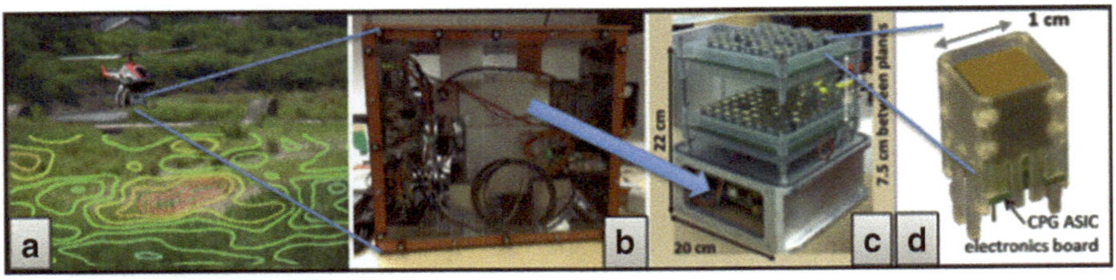

Fig. 18.3 (**a**) The High-Efficiency Multimode Imager HEMI during operation in the Namie-Machi area. (**b**) The internal components of the environmental housing for HEMI providing autonomous operation of the system during operation. (**c**) The HEMI instrument composed of 96 coplanar grid (CPG) CdZnTe (CZT) detectors arranged in two layers enabling coded aperture gamma-ray imaging and Compton imaging simultaneously. (**d**) One of the 96 1 cm^3 CPG CZT detector elements

of ground and aerial vehicles such as trucks and helicopters. In addition, advanced models have been developed to estimate gamma-ray intensities and doses on the ground over different spatial and temporal scales. Figure 18.3 illustrates the High-Efficiency Multimode Imager (HEMI) that was developed at LBNL mounted on a RMAX helicopter provided by JAEA. Measurements were performed close to Namie-machi in the evacuated area North-West of the Dai-ichi nuclear power plant [12, 13]. HEMI can be operated as handheld gamma-ray detector and imager or as shown on an unmanned aerial platform. The dimensions and weight of HEMI is well matched to the operation on a RMAX helicopter.

18.2.2.2 Characterization and Prediction of Contaminant Fluxes in the Environment

One important aspect of efficiently responding to radiological and nuclear accidents is to understand and predict the long-term transport of radionuclides within and between different environmental compartments, such as farmland and forests, water bodies such as rivers, lakes, and reservoirs, and soil and groundwater. Predictive studies are needed to evaluate and compare the effectiveness of active and passive remediation options. Figure 18.4 illustrates the challenges associated with the very complex environmental transport in the Fukushima prefecture with a combination of forested and mountainous terrains with different solid and water surfaces, and significant temporal changes due to seasonal variations. LBNL has developed advanced mechanistic models describing the transport of cesium and other contaminants from each of the compartments of interest [14–17]. These models are being extended across a wide range of scales, from the pore scale to the larger sub-watershed and watershed scales, to develop a hierarchical multi-scale modeling framework that is integrated closely with appropriate scale observations and characterization. The modeling framework will be applied to and tested against data from representative test regions in the Ogi-na-Sawa basin near the Fukushima site, where detailed

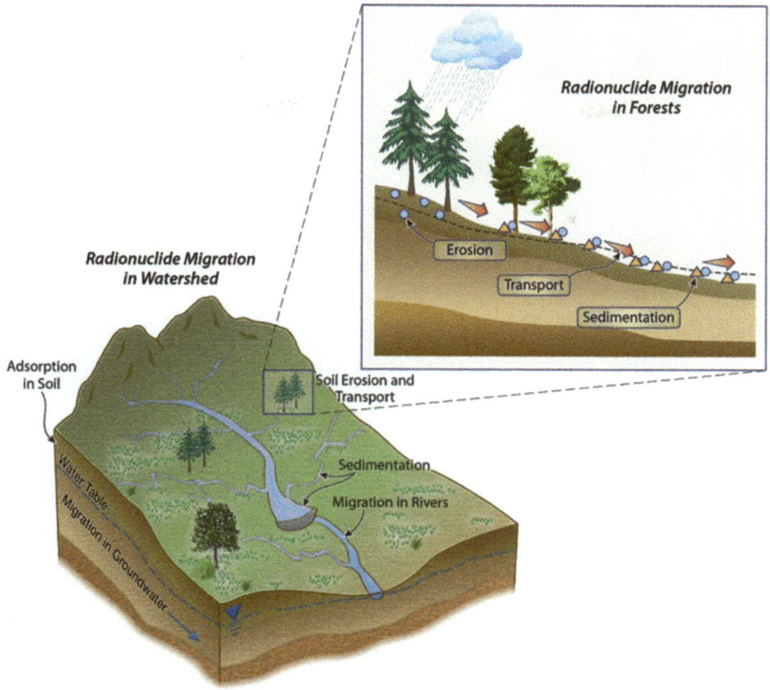

Fig. 18.4 Illustration of examples of transport mechanisms required to describe and predict the transport of radioisotopes in the complex environment of Fukushima. Of particular interest for long-term predictive models is the forest as it captures most of the stored cesium contamination in the Fukushima Prefecture

characterization and monitoring is being conducted by JAEA to better understand cesium fluxes from forests into rivers and reservoirs. The intermediate scale models, which are designed to capture the integrated flux of cesium in both the solid and the aqueous phase, will be based on mechanistic representations of relevant processes, including multicomponent geochemistry, plant litter accumulation and subsequent decay, transient overland flow, and transient groundwater flow.

18.2.2.3 Alleviating the Health Impact of Radiation Exposure

Following a radiological or nuclear accident, a large population may be at risk for external exposure and internal contamination. In order to respond quickly and effectively to releases of radiological materials and to guide long-term remediation efforts, technologies are required that can assess the health impact on individuals, predetermine sensitive populations, and design guidelines for triage and mitigation to implement a rapid response. Ionizing radiation leads to DNA damage that can lead to acute toxicity or long-term risk such as cancer. The understanding of health effects associated with specific levels of radiation, and the development and acquisition of prophylactic and post-exposure medical agents that protect against the debilitating effects of internalized radionuclides provide a needed safety measure for individuals.

A high throughput imaging approach to quantify DNA damage from blood samples has been recently developed at LBNL. It is becoming possible to apply this technology to characterize and predict DNA damage and repair from chronic exposure to internal contamination and to external ionizing radiation [18, 19]. In addition, novel chelating agents have been recently developed at LBNL for post-exposure decorporation that have demonstrated strong prophylactic potential and will be tested as therapeutics for contaminated individuals. Developing and delivering protective prophylactic and post-exposure medical agents are essential for mitigating radiation health effects in individuals [20, 21].

18.2.2.4 Mitigation and Remediation – Classification and Evaluation of Chemical Treatment of Fukushima Soils

Understanding the chemical speciation of radionuclide contaminants in the environment is a key factor for proper risk assessment, prioritization, and development of efficient and effective remediation processes. Chemical speciation is necessary for the implementation of chemical processes, intermediate waste storage options, and waste reduction technologies. The classification and remediation of materials containing cesium the Fukushima prefecture can be significantly accelerated by the determination of cesium speciation in soils by soft X-ray synchrotron radiation spectromicroscopy using a scanning transmission X-ray microscope (STXM) at the Advanced Light Source (ALS) [22]. These efforts include the speciation of radionuclides at surfaces and in soils, water, and biological materials under in-situ conditions at relevant concentrations and length scales, ranging from nm to cm. Illustrated in Fig. 18.5 is the use of the scanning transmission X-ray microscope to determine the spatially-resolved speciation of cesium "hot spots" in clay particles (top) by X-ray absorption spectroscopy (bottom).

18.3 Status and Path Forward

The aforementioned outreach and research activities will become the central pillars for the proposed Institute for Resilient Communities. Based on these established activities we will expand our outreach and research activities locally in Berkeley and with our research and community partners and organizations in Japan. Reflecting the need to work with local communities we are pursuing partnerships with cities such as Berkeley in the U.S. and Koriyama in the Fukushima Prefecture in Japan.

Figure 18.6 captures the internal and external relationships of the ongoing science and research components. The four research threads are being pursued

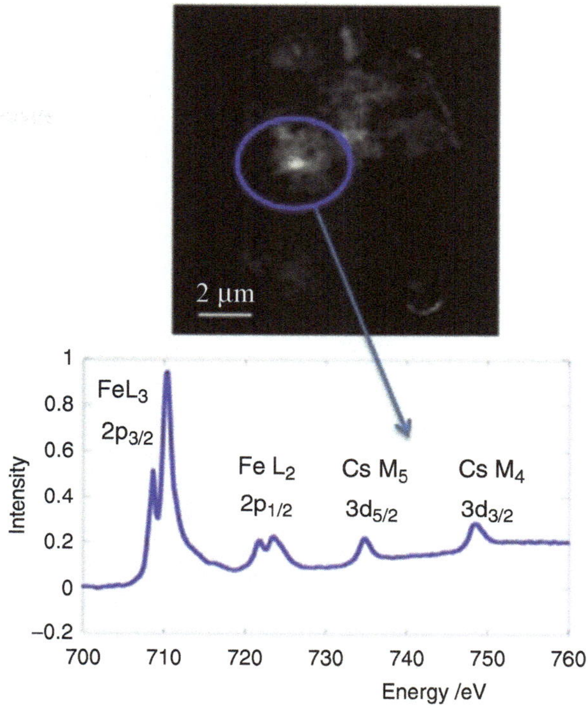

Fig. 18.5 Illustration of the use of the scanning transmission X-ray microscope to determine the spatially-resolved speciation of cesium "hot spots" in clay particles (*top*) by X-ray absorption spectroscopy (*bottom*)

by scientists of four different divisions of LBNL and the Department of Nuclear Engineering at UC Berkeley.

Complementary to the research, we will continue Radwatch with its near-realtime air monitoring and measurements of environmental and food samples as well as with Kelpwatch as part of the outreach and educational efforts. As before and reflecting the importance of transparency in any of our activities to maintain public trust, we will publish all our measurements, procedures, and findings. While the focus to-date was on gamma-ray spectroscopy that is sensitive to radioisotopes such as Cs-134 or Cs-137, we will expand our measurements to alpha spectroscopy which will allow us to measure radioisotopes such as Po-210 which are also naturally occurring, however, can also be misused as highly effective poison [23].

We are developing and installing a radiation dosimeter network, first across UC Berkeley campus and then across local schools. In parallel, we will make dosimeters available to these schools and will describe science projects that can be performed by the students. These projects have two objectives: (1) Allow the students to "see" radiation in our environment and to learn important properties of radiation, e.g. the fact that it varies spatially and temporally or that it can be shielded or reduced by increasing the distance; (2) Enable a better understanding and appreciation of fundamental concepts in science and engineering such as uncertainties associated with observations and measurements, statistics and probability, and ultimately risk. Particularly the concept of risk is becoming ever more important in our modern and

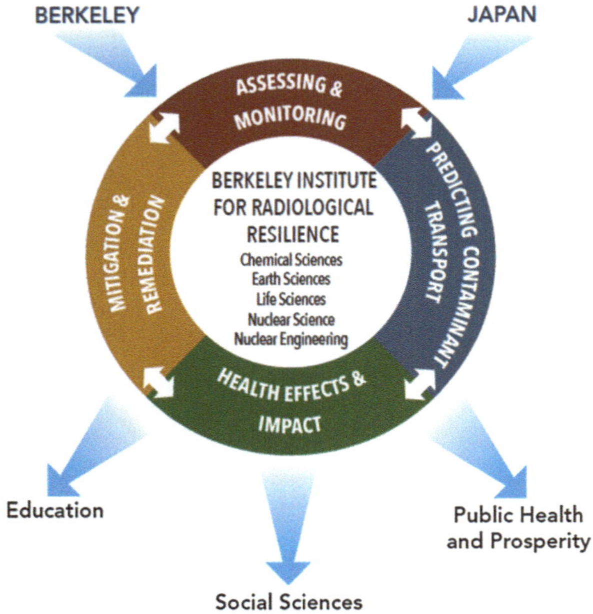

Fig. 18.6 Illustration of the research and science components of the proposed Berkeley Institute of Radiological Resilience that are based on already ongoing or planned collaborations with Japanese partners and its envisioned impact on education, social sciences, and public health

technological driven global society and therefore needs to be better understood by the public. The first objective addresses specifically the fear of radiation in the public as it can not be recognized with human senses. The second objective addresses the need to enhance more broadly the science and technology literacy of citizens.

18.4 Summary

Recent events associated with the releases of radioactive materials and the recognition of the possibility of events that are associated with the release of radioactive materials to happen in the future represent major challenges for advanced and global societies. Radiological or nuclear events due to accidents or the misuse of materials have and will have an enormous socio-economical and political impact on local and global communities. While it is possible that such an event may have significant health effects due to radiation, the psychological impact will be substantial, as observed in Japan. While it is paramount to enhance the safety and security of the currently operating and future nuclear power plants, it is also critical to enhance means in responding and recovering from a possible event to minimize the impact of such events, i.e. to increase the resilience to such events.

The Institute for Resilient Communities addresses this need by combining natural and social sciences, technology and engineering, education and outreach and involves local communities, all in a global context. It addresses the need to improve the scientific understanding of cause and impact of such events. Specifically, the transport and impact of the released radioactive materials in the geo-sphere and the bio-sphere, including humans needs to be better understood. Technologies are being developed to provide the necessary data and information to enable effective means to respond and recover from an event. This includes radiation and bio-sensor technologies as well as modeling and communication tools. The education and outreach aspect aims at minimizing the psychological effects through a better-informed public. While the focus will be on radiation, the goal is to establish programs to enhance science and technological literacy more broadly, including basic concepts in science and engineering. Data will be collected and made available to recognize and appreciate the world we are living in, particularly the world we can not see or feel. Local communities will be involved to effectively introduce these concepts to the public and into schools. By providing such a framework, the Institute for Resilient Communities will become a trusted resource to the public, media as well as decision makers, essential in the response to a radiological or nuclear event.

Acknowledgments I thank the Berkeley radiological resilience team led by Rebecca Abergel, Jens Birkholzer, Sylvain Costes, and David Shuh from LBNL and Joonhong Ahn from UC Berkeley for their essential contributions to the development of the concept and the realization of radiological resilience. Hiroyuki Takahashi from Univ. Tokyo and Tatsuo Torii from JAEA were critical in enabling our measurements in Japan. Support from Mark Amman, Dan Chivers, Michelle Galloway, Andy Haefner, John Kua, and Paul Luke made the development and the measurements with HEMI possible. Ryan Pavlovsky effectively led the Radwatch activities. We acknowledge funding from LBNL, the Domestic Nuclear Detection Office, and Furukawa LLC.

References

1. van Deventer E (2014) WHO's health risk assessment based on a preliminary dose estimation of the Fukushima Daiichi NPP accident. In: Third international expert symposium in Fukushima: beyond radiation and health risk – toward resilience and recovery, Fukushima City, Japan, 8–9 September 2014
2. Committee on Technological Literacy, National Academy of Engineering and National Research Council (2002) Why all Americans need to know more about technology. Edited by Pearson G, Young AT. National Academy Press, Washington, DC. Available at http://search.nap.edu/books/0309082625/html
3. Miller HI (2014) Flu is a far greater menace than ebola in US, Forbes Magazine. http://www.forbes.com/sites/henrymiller/2014/10/29/ebola-shmebola-flu-is-a-far-greater-threat/. 29 Oct 2014
4. Center for Disease Control and Prevention/National Vital Statistics System. Influenza/Pneumonia mortality by State: 2012. http://www.cdc.gov/nchs/pressroom/states/FLU_PNEUMONIA_STATE_2012.pdf
5. http://radwatch.berkeley.edu, http://kelpwatch.berkeley.edu
6. Fisher NS, Beaugelin-Seiller K, Hinton TG, Baumann Z, Madigan DJ, Garnier-Laplace J (2013) Evaluation of radiation doses and associated risk from the Fukushima nuclear accident to marine biota and human consumers of seafood. Proc Natl Acad Sci 110(26):10670–10675. http://www.pnas.org/content/110/26/10670.full.pdf
7. Madigan DJ, Baumann Z, Fisher NS (2012) Pacific bluefin tuna transport Fukushima-derived radionuclides from Japan to California. Proc Natl Acad Sci 109(24):9483–9486. http://www.pnas.org/content/109/24/9483.full.pdf

8. Madigan DJ, Baumann Z, Snodgrass OE, Ergl HA, Dewar H, Fisher NS (2013) Radiocesium in Pacific bluefin tuna thunnus orientalis in 2012 validates new tracer technique. Environ Sci Technol 47:2287–2294. http://dx.doi.org/10.1021/es4002423
9. Woods Hole Oceanographic Institution (2014) Fukushima radioactivity detected off west coast, Nov 2014. http://www.whoi.edu
10. Nuclear Regulatory Agency, Japan (2015) Distribution map of seawater radioactivity around TEPCO Fukushima Dai-ichi NPP. In: Monitoring information of environmental radioactivity levels, Jan 2015. http://radioactivity.nsr.go.jp/en/contents/10000/9873/24/349_20150611.pdf
11. Homma T (2014) Radiological situations after the Fukushima accident and arrangements for preparing long-term recovery operations. In: Third international expert symposium in Fukushima: beyond radiation and health risk – toward resilience and recovery, Fukushima City, Japan, 8–9 September 2014
12. Amman M, Luke PN, Lee JS, Mihailescu L, Vetter K, Zoglauer A (2009) Detector module development for the High-Efficiency Multimode Imager. IEEE TNS 10(1109):981–985
13. Zoglauer A, Galloway M, Amman M, Luke PN, Boggs SE (2010) Aerial standoff detection with High-Efficiency Multimode Imager HEMI. IEEE TNS 10(1109):566–570
14. Steefel CI, Carroll S, Zhao P, Roberts S (2003) Cesium migration in Hanford sediment: a multi-site cation exchange model based on laboratory transport experiments. J Contam Hydrol 67:219–246
15. Li L, Gawande N, Kowalsky MB, Steefel C, Hubbard SS (2011) Physicochemical heterogeneity controls on uranium bioreduction rates at the field scale. Environ Sci Technol 45(23):9959–9966. doi:10.1021/es201111y
16. Regnier P, Wollast R, Steefel CI (1997) Long-term fluxes of reactive species in macrotidal estuaries: estimates from a fully transient, multicomponent reaction-transport model. Mar Chem 58:127–145
17. Bea S, Wainwright H, Spycher N, Faybishenko B, Hubbard SS, Denham ME (2013) Identifying key controls on the behavior of an acidic-U(VI) plume at the Savannah River Site using reactive transport modeling. J Contam Hydrol 151:34–54
18. Vadhavkar N, Pham C, Georgescu W, Deschamps W, Heuskin AC, Tang J, Costes SV (2014) Combinatorial DNA damage pairing model based on X-ray-induced foci predicts the dose and LET dependence of cell death in human breast cells. Radiat Res. doi:10.1667/RR13792.1
19. Neumaier T, Bissell MJ, Thalhammer S, Costes SV (2012) Evidence for formation of DNA repair centers and dose-response non-linearity in human cells. Proc Natl Acad Sci U S A 109(2):443–448. Epub 2011 Dec 19
20. Bunin DI, Chang PY, Doppalapudi RS, Riccio ES, An DD, Jarvis EE, Kullgren B, Abergel RJ (2013) Dose-dependent efficacy and safety toxicology of hydroxypyridinonate actinide decorporation agents in rodents: towards a safe and effective human dosing regimen. Radiat Res 179:171–182
21. Jarvis EE, An DD, Kullgren B, Abergel RJ (2012) Significance of single variables in defining adequate animal models to assess the efficacy of new radionuclide decorporation agents: using the contamination dose as an example. Drug Dev Res 73:281–289
22. Bluhm H, Andersson K, Araki T, Benzerara K, Brown GE Jr, Dynes JJ et al (2006) Soft X-ray microscopy and spectroscopy at the molecular environmental science beamline of the advanced light source. J Electron Spectrosc Relat Phenom 150:86–104
23. McFee RB, Leikin JB (2009) Death by polonium-210: lessons learned from the murder of former Soviet spy Alexander Litvinenko. Semin Diagn Pathol 26(1):61–67

Chapter 19
Societal Consequences of Nuclear Accidents

Astrid Liland

Abstract Both accidents at nuclear installations and terrorist attacks can cause severe radioactive contamination over large areas. The severity and long persistence of radioactive contamination challenges the affected communities in many ways. It is not just a question of dose – it affects environment, economy, production, living conditions and health. It is thus a societal problem and the management strategy after a nuclear accident needs to take account of social, ethical and economic consequences along with the radiation impact. In addition, people generally have little knowledge about radioactivity and radiation and are not prepared to tackle a contamination event. The fear of something that is invisible and at the same time ubiquitous and invasive in people's life, can be very difficult for people to handle. Situations where large areas within a country would be contaminated for a decade or more would represent a huge challenge to any country. There is a range of possible countermeasures that could be implemented after a nuclear or radiological accident, directed both at the population and at the production of food, feed and goods. However, actions in the recovery phase need to be chosen with care, taking account of the wider societal aspects and preferably in elaboration with people from local, regional and national levels. The goal of the remediation strategies would be a return to a normal situation which is not a pre-accident situation, but a situation where people can live and produce in a contaminated area with acceptable risk and living conditions due to the implementation of mitigating actions.

19.1 Acknowledging Risk

We all know that nuclear accidents happen, although not frequently. The Windscale accident in 1957 [5] and the Three Mile Island accident in 1979 are both examples of accidents with limited off-site consequences, while the Chernobyl accident in 1986 was a catastrophic event with large-spread contamination over Europe and

A. Liland (✉)
Centre for Environmental Radioactivity and Norwegian Radiation Protection Authority, Oslo, Norway
e-mail: Astrid.Liland@nrpa.no

S. Apikyan, D. Diamond (eds.), *Nuclear Terrorism and National Preparedness*, NATO Science for Peace and Security Series B: Physics and Biophysics, DOI 10.1007/978-94-017-9891-4_19

decades of problems. Still, it was not until the Fukushima accident in 2011 that nuclear power countries acknowledged that severe accidents could happen in well developed countries with newer reactor designs. The risk of natural disasters, multiple-unit failure, complete electricity black-out and long term releases had to be acknowledged as possible, worst-case scenarios. It changed the view on necessary safety, security and emergency preparedness plans across the world.

Since 9/11 2001, the focus on preventing terrorist attacks has also been prominent, including considerations of malevolent acts including radioactive material. So far there has not been any terrorist attacks involving radiation, but we still need to prepare for this unlikely, yet not impossible, event. What would have happened if the planes that crashed into the Twin Towers on 9/11 also carried easily dispersible radioactive material?

In Norway, the sole terrorist Anders Behring Breivik performed an attack on 22 July 2011 [4]. He blew up six governmental buildings using a car bomb made from fertilizers, killing eight people and injuring many more. He then drove to a political youth camp at Utøya outside Oslo where he shot and killed 69 people, mostly teenagers, and injured 66 before the police managed to bring him down. Just before his attack he had published a Manifesto "A European Declaration of Independence" more than 1,000 pages long. It contained around 70 pages where he described the use of tactical nuclear weapons, attacks on nuclear facilities, use of Radiological Dispersion Devices and CBRN-contaminated ammunition. The Oslo police bomb squad sent in robots with radiation detectors in the damaged governmental building to establish whether radiation was involved or not. At the Oslo University Hospital, where the injured teenagers from Utøya were admitted, the retrieved ammunition was scanned for radioactivity.

A commission was formed to investigate the shortcomings of the Norwegian first responders and the emergency response units/agencies during this attack. In their Official Norwegian Report [4] they made some conclusions that could be valid also for other scenarios and countries:

- The ability to acknowledge risk and learn from exercises has not been sufficient;
- The ability to implement decisions that have been made, and to use the plans that have been developed, has been ineffectual;
- The ability to coordinate and work together has been deficient;
- The potential inherent in information and communications technology has not been exploited well enough; and
- Leadership's willingness and ability to clarify responsibility, set goals and adopt measures to achieve results have been insufficient.

19.2 Moving into the Long Term

The conclusions above are related to the actions in the emergency phase. Plans are usually in place for early phase response actions, but if the plans are not well known, shared by relevant actors and exercised in peace time, they have limited value when an accident happens. Moreover, many countries lack plans for response in the long term for nuclear accidents or malevolent acts that create large areas contaminated by radioactive material. For a country to be resilient to these kinds of events, you need to plan for the long term, too. From experience we know that the contamination could be a problem for decades.

Let's take Norway as an example. Norway is situated more than 1,000 km from the Chernobyl NPP, yet it was the country outside the former Soviet Union that was most heavily affected by the Chernobyl fallout. The fallout areas coincided with important pasture areas for sheep, goats, cows and reindeer [3]. After a period of denial, then confusion, a management system was put in place by the authorities from summer 1986. Countermeasures the first year were dominated by monitoring, food bans and setting food intervention levels. When the situation was better mapped and the authorities realized that the impacts could last for many years, new countermeasures were developed during the next couple of years. We are still using countermeasures in Norway today to produce meat and milk that comply with the food intervention levels and we offer whole body counting for the population groups at risk. The most important countermeasures are:

- Elevated food intervention levels for reindeer, game and freshwater fish ('minor foodstuffs')
- Monitoring of radiocaesium in animals before slaughter ('live monitoring');
- Clean feeding of animals before slaughter;
- Cesium binder (Prussian blue) in concentrates, salt licks and rumen boli to prevent absorption of ingested radiocaesium in the animals;
- Change of slaughter time (in reindeer husbandry); and
- Dietary advices and monitoring of internal contamination.

They were developed as a cooperation between authorities, agricultural experts, farmers and local administrations during 1986–1988. For more details, see [2, 3]. Figure 19.1 shows the time development of Cs-137 in goat's milk from 1988 to 2014 for a herd during summer grazing. It shows a similar trend as for other areas that produce meat and milk: the levels are high the first years then decreasing over time due to natural decay. In years abundant in mushrooms, you will see peak values. Mushrooms are more contaminated than green fodder and are also preferred by the grazing animals.

The ecological half-life is often a two-component one with a rapid initial decay and a much slower second component. This is shown for many different herds in Norway in Table 19.1. Figure 19.2a shows the equipment used for measuring live animals. It consists of a Canberra Inspector 1000 with a 3" NaI detector in a plastic capping. A specific application was developed by NRPA and Canberra so that the device will display the amount of Cs-137 in the meat directly in Bq/kg. Figure 19.2b shows the position of the detector on the sheep during measurement. Live monitoring of animals are used to determine whether the animals can be slaughtered or if a period of clean feeding needs to be imposed for the herd. The number of weeks with clean feeding varies from 1 to 8, usually in the lower range nowadays. Clean feeding is based on the principle that the radiocaesium will be

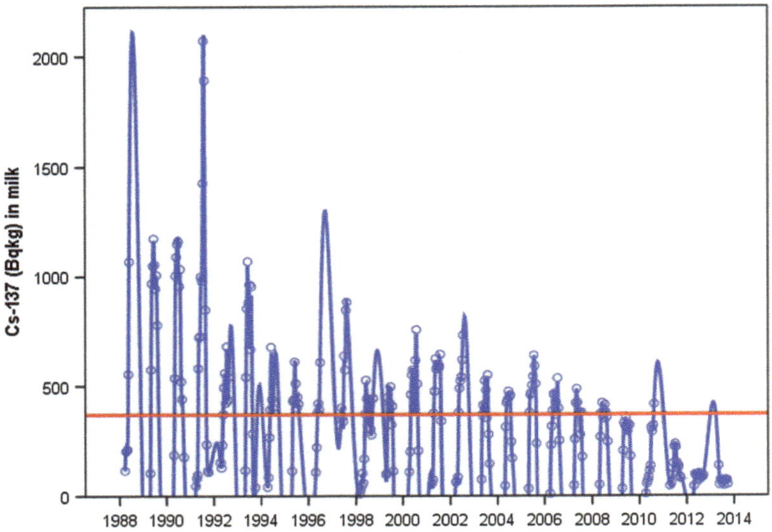

Fig. 19.1 Measurement of goat milk during the summer. The *red line* represents the food intervention level for milk of 370 Bq/L [1]

Table 19.1 Ecological half-lives of ^{137}Cs derived from monitoring of milk and live monitoring of sheep [1]

Product	County	Municipality	Observation period	Ecological half-life (a)	Ecological half-life (a)	N
Sheep muscle	Oppland	Vestre Slidre	1991–2014	0,8 ± 0,3	20,8 ± 3,4	2,287
Goat milk	Buskerud	Ål	2000–2013	∞	∞	146
Goat milk	Buskerud	Ål	1989–2006	∞	∞	132
Goat milk	Oppland	Øystre Slidre	1988–2013	1.4 ± 2.0	12.9 ± 3.1	259
Cow milk	Oppland	Vang	1996–2013	2.0 ± 1.1	12.5 ± 3.3	557
Cow milk	Oppland	Øystre Slidre	1990–2013	0.5 ± 0.3	11.0 ± 0.9	380
Cow milk	Oppland	Øystre Slidre	1998–2013	1.5 ± 0.9	34.2 ± 12.4	186
Cow milk	Nordland	Brønnøy	1998–2013	0.2 ± 1.4	52.5 ± 41.2	94
Cow milk	Nordland	Vega	1991–2013	0.3 ± 0.1	7.7 ± 0.4	134
Cow milk	Nordland	Vevelstad	1991–2013	2.9 ± 3.5	∞	72
Cow milk	Nordland	Vevelstad	1991–2013	5.2 ± 5.5	∞	148
Goat milk	Nord-Trøndelag	Namsskogan	1989–2012	2.8 ± 3.1	5.9 ± 3.4	212
Goat milk	Nord-Trøndelag	Røyrvik	1998–2006	0.1 ± ∞	17.5 ± 2.4	222

excreted from the animals according to the biological half-life which is around 3 weeks for sheep meat. Live monitoring and clean feeding is still extensively used in Norway today. It is clear that the consequences from the Chernobyl accident are long lasting and we foresee the need for countermeasures for another decade at least.

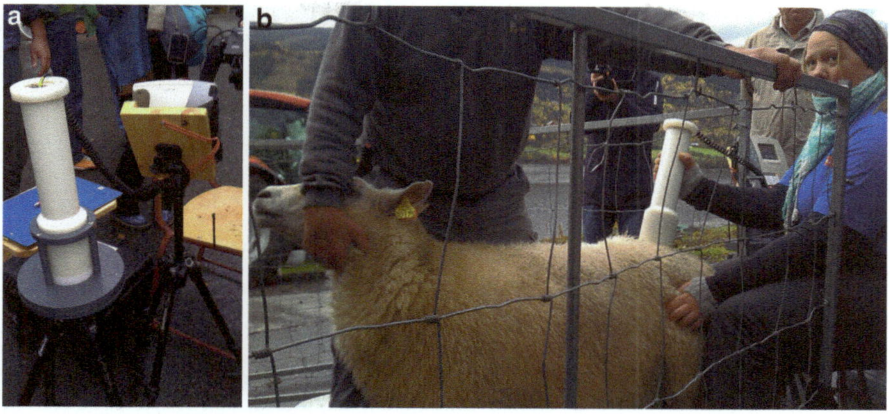

Fig. 19.2 (**a**, *left*): Equipment for live monitoring and (**b**, *right*): Position of the detector while measuring sheep (Photos: Astrid Liland, NRPA)

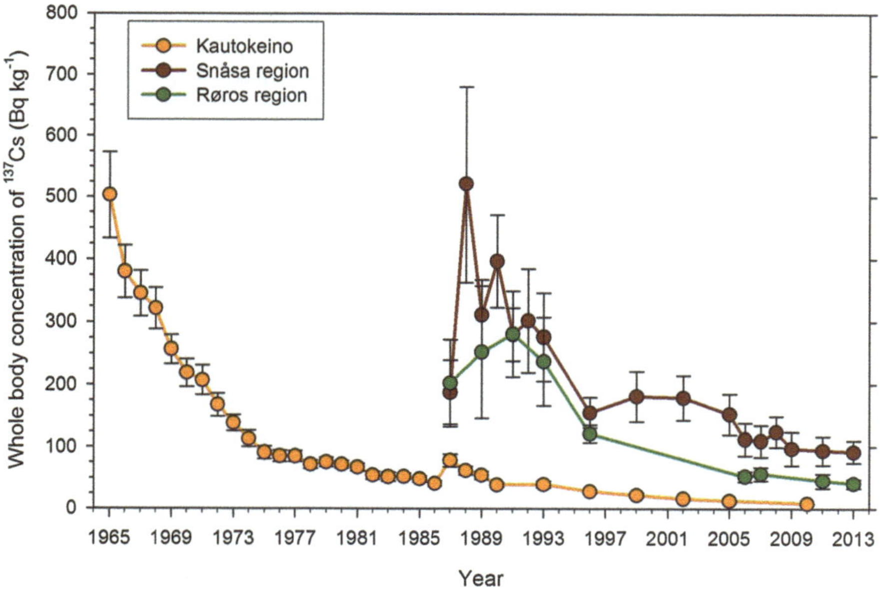

Fig. 19.3 Group average WBC values for different Sami populations in Bq of ^{137}Cs per kg of bodyweight from 1965 to 2013. A value of 400 Bq/kg chronic exposure corresponds to approximately 1 mSv/y effective internal dose

NRPA has also offered whole body counting (WBC) to populations at risk. Figure 19.3 shows the measurement results for three groups of Sami people, the indigenous people of Norway. Due to a high consumption of reindeer meat these groups are the ones most exposed in Norway both historically and today. The orange line shows the values for Northern Sami from Kautokeino impacted from the atmospheric nuclear weapons testing. The red and green lines show the values for Southern Sami impacted by the Chernobyl accident. The values are given as average group values in Bq per kg bodyweight.

According to [6], the use of various countermeasures and dietary advice has reduced the exposure to the Southern Sami population group at Snåsa with approximately 70 %.

19.3 Nuclear Fallout – A Long-Term Societal Challenge

From the examples shown it is clear that radioactive fallout is a long term problem and it affects all parts of the society: health, environment, economy, production, living conditions – it is a complex situation. Let's take the Fukushima accident as an example. Large areas were contaminated and more than 150,000 people evacuated. Most of them are still evacuated 4 years after the accident. Industrial and agricultural production was stopped in evacuated zones and is only now slowly been taken up in certain areas where they are allowed to go back. There was, and still is, a ban on fisheries in the 20 km zone outside the Fukushima NPP. There are still restrictions as to which species can be harvested and used for human consumption. Only a fraction of the earlier fishery industry in Fukushima is still operational. Agricultural produce was contaminated in large areas. Even if the situation has improved and decontamination of agricultural fields has been performed in many areas, the producers find it difficult to sell their produce both within and outside of Fukushima.

The reality for people, in the long term, is what's going on in their hometown – not in a national crisis centre. For the people of Fukushima the reality is that many people are still evacuated, the communities and families have been dispersed. Their traditions and culture are discontinued due to this. Many people are reluctant to move back even if radiation levels are now below 1 mSv/y. It is difficult to be farmers and fishermen, many businesses have been disrupted. The compensation scheme creates suspicion among neighbours, induces passivity and is a constraint in the return to earlier evacuated areas. People who want to move back have concern over the lacking infrastructure, future schools and health services, possible livelihoods and the demographic shift towards more elderly people. They also experience stigma and discrimination by others just for being from Fukushima.

Japan had not made response plans beyond the evacuation phase. The rehabilitation work has been very slow due to this and other aspects of Japanese society and politics. Countries need to have response plans for large scale contamination and long term impacts. There are important questions to ask relative to your country's emergency response plans. Have you planned also for the long term? Once the fallout is a fact and people around the NPP have been evacuated, then what? How will you work for the rehabilitation of the areas and a return to normality?

There are many tools available for the response phase of a nuclear accident, also for the long term. There are decision support systems, handbooks, guidelines, regulatory frameworks etc. But the important questions are:

- Do you study them in peace time?
- Have you adapted them to your country/region?
- Are they well known to all actors in emergency preparedness and response in your country?
- Do you use them in exercises?

You can have the best plan in the world, but if it is tucked away in a drawer, it is of no use!

You need to plan also for the long term and ask yourself important questions like:

- Will the evacuation be temporary?
- Which mitigating actions will you implement?
- Who will perform the actions?
- Where to dispose of all the waste generated?
- How can you engage the municipalities in emergency preparedness planning and exercises to create resilience?

19.4 Zoning in Fukushima – Is Evacuation Really Evacuation?

The contaminated areas in Fukushima have been divided into three zones, see Fig. 19.4. The most contaminated zone (red) have external doserates >50 mSv/y. It has been termed the 'Difficult to return zone' by Japanese authorities who have indicated that people may never be allowed to return to this zone in their lifetime. Almost 25,000 people are still evacuated from this zone alone.

The intermediate contaminated zone (orange) has an external radiation level between 20 and 50 mSv/y. People are allowed to visit the area, but not to resettle for the moment. More than 23,000 people are still evacuated from this area where the authorities will perform the decontamination work.

The least contaminated area (green) has less than 20 mSv/y external exposure. Local authorities have/will perform the decontamination in this area. Some have been allowed to move back since April 2014, but more than 30,000 people are still evacuated from this area.

And we are now 4 years after the accident. In some areas we must assume that it will be a permanent relocation rather than an evacuation. Also, the longer people stay evacuated, the less eager they are to return to their original home place, in particular families with school children who establish a new life in a different community.

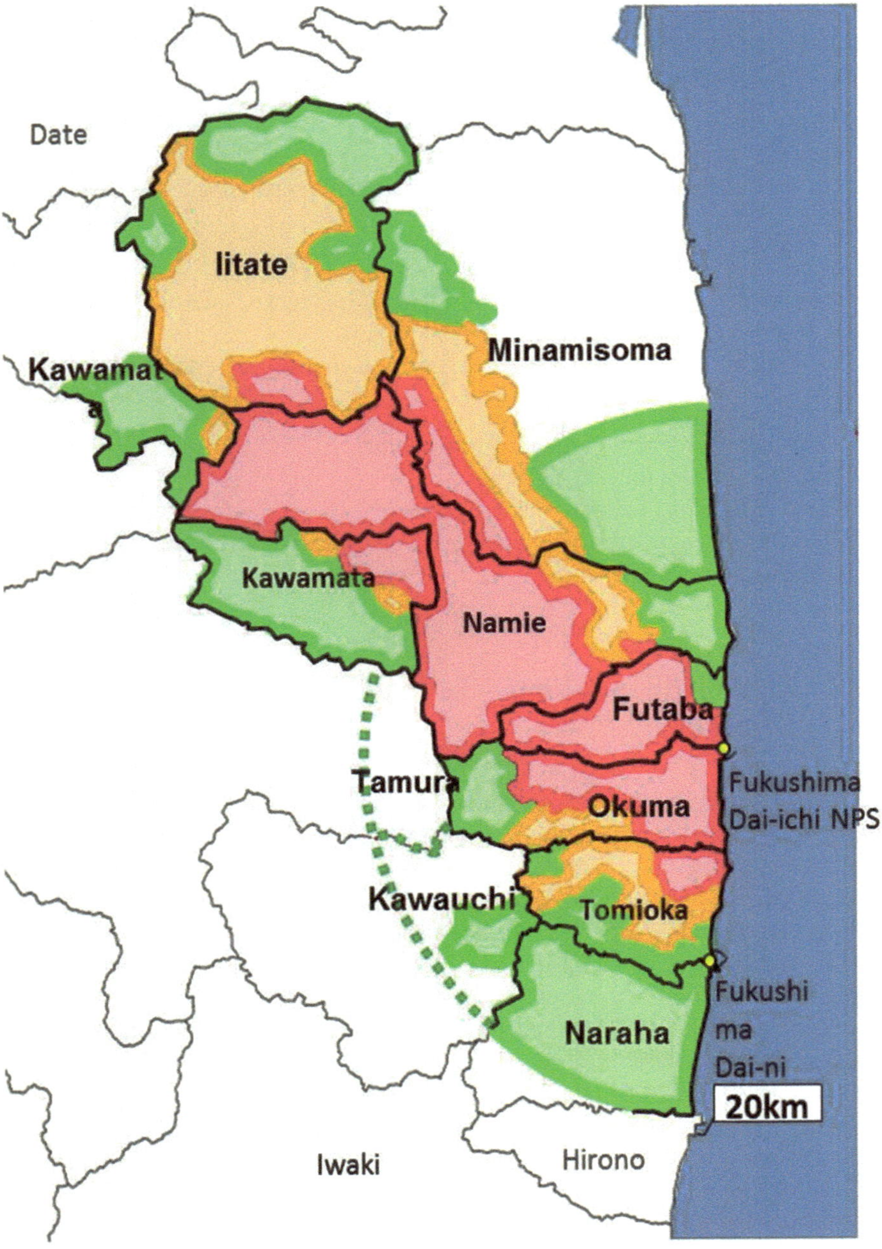

Fig. 19.4 Current zoning in the Fukushima district (Reprinted with permission from Support Team for Residents Affected by Nuclear Incidents, Cabinet Office)

Decontamination of areas creates a lot of waste that need to be disposed of. Have your community made plans for storage of this kind of low level waste in large quantities?

Another important question is who will perform the mitigation action. Often, the impacts are of a durable character and the mitigating actions need to be performed in the districts by local people. It is thus imperative to involve local municipalities in the development of rehabilitation plans. The aim is to restore normal living conditions for people in the contaminated territories. This is not a pre-accident situation, but a new normality where people are empowered to continue to live and produce in the affected areas with acceptable risk. The government should plan how to cooperate between local, regional and national level in the recovery and long term management phase. Local administrations and the public lacks knowledge of how to handle the situation and don't always understand the information given. At the same time, most people are rational and willing to help out, but they need expert assistance. Help to self-help is a much better solution to revive a society than to just offer monetary compensation and tell people to wait until the government will come and 'make it right'. In this way you will also make use of the important resources that lies in local communities and their inhabitants.

19.5 The Example of Food Contamination and Food Bans

In the aftermath of a nuclear accident, food bans will often be necessary or even mandatory to ensure food safety. It is efficient for removing contaminated food from the market in a situation of confusion and limited measurement data. But in the long term, food bans have many side effects:

- Generate a lot of waste
- Are costly (compensation to producers, measurement campaigns, waste disposal, lost market value)
- Stop in production changes the cultivated landscape
- Stop in harvesting changes the ecological balance (eg. wild boar in Fukushima Prefecture)
- Loss of production knowledge
- Import/export disturbed
- Food shortage?
- Malnutrition?

Last, but not least, it is very unsatisfactory for the farmers, fishermen and hunters.

It is important to understand the socio-ethical aspects of food production. There is a personal value in providing quality food through agriculture, fishing and hunting. It is part of culture heritage: continue the traditions of former generations, use of the land and value of regional produce like wine, cheese, rice or other. If you are not allowed to continue your traditional agricultural production, for instance, the fields will be regrown and the appearance of the traditional landscape will change. Imagine that the French people would no longer be allowed to produce cheese and wine. Would they still be French? Or if the wine producers in Napa Valley in California were not allowed to continue their production, what would this mean for the identity of inhabitants in the valley? Or if people felt uneasy about going to the beach in California due to (rumor of) radiation, what would this means for the feeling of being Californian?

Food bans and the destruction of foodstuffs are very unrewarding for the farmers, fishermen, wine producers etc. They take their pride in producing quality food and the authorities should assist them in this task also in the case of radioactive fallout by providing local monitoring stations. This has proven valuable in both Norway and Belarus after the Chernobyl accident [2]. Local measurements performed by known staff are more trusted than country average values provided by central agencies. In many cases, government independent laboratories are also valuable. In particular when government measurements are not trusted or not sufficient, the additional measurement capacity of independent laboratories can be highly valued by the public. It can also help in regaining the trust of consumers who are dependent on good information on, and understanding of, measurement results to buy food from affected areas.

It is clear that, depending on contamination level, land use, environmental factors etc., the producers could be facing decades of problems. Food bans and production prohibition are not viable in the long term, thus a countermeasure strategy should be elaborated with the producers. Experts are needed to assist local communities in developing strategies and local monitoring stations. Active participation in mitigating actions and access to local monitoring stations empowers the producers and the public and leads to less psycho-social stress [2, 3]. It is also my personal belief that compensation should be paid only to cover the costs associated with implementation of countermeasures and not just for being a victim of radioactive contamination. (Of course here, compensation to evacuees is of a different, and necessary, character). Compensation for just living in a contaminated territory may induce passivity and enhance the feeling of being a 'victim', as well as contribute to the stigma and discrimination of people in an affected area. The issue of monetary compensation is very delicate and must be exercised with care.

For all affected people, the involvement in rehabilitation strategies contributes to autonomy, freedom/liberation and equity – key elements in the philosophy of science. Science is not only facts and figures; science also has a valuation of statements and facts. Science should be at the service of humanity and need to take a larger reality into account besides the natural sciences. Such an involvement would contribute to a collective effort:

- Allowing new views and perspectives, and creative problem solving; and
- Making use of all the human and intellectual resources available.

Common evaluation and problem solving by evacuees, residents, experts and authorities (co-expertise) in the various communities should be supported. Places, processes and plans for such cooperation need a national framework and should preferably be planned for in advance.

19.6 Use MUM, Not DAD in Engagement with the Public

The authorities and experts need to engage with the public in the following way:

Meet them, **U**nderstand their challenges and **M**odify plans for recovery and rehabilitation to meet the inhabitants' needs (MUM, according to ICRP). The old strategy of **D**ecide, **A**nnounce, **D**efend (DAD) is not a good strategy for successful cooperation with the affected communities. In short: Make plans with the people, not for the people.

19.7 Conclusions

It is clear from the experience from former accidents that governments need to acknowledge the possible severity and long duration of contamination from nuclear accidents. The plans for the recovery and late phase must cover, and be shared by, all sectors and levels of the society if they are to be successful. Authorities and local administrations alike must familiarise themselves with existing tools and adapt them to national and regional needs. Involvement of the affected communities should be conducted through MUM, not DAD. Exercises involving all sectors and levels are essential for building societal resilience towards RN emergencies. And a last fact not treated above, but yet very important: Today information moves fast! And through many channels. Authorities are wise to prepare general information in advance that can easily be adapted for a given accident situation. The information must be distributed fast and you must be available through all channels. You want to be the preferred source of information! Remember, your success in emergency management will only go down in history as a success if this view is shared by your citizens.

Acknowledgement This work has been supported by the Research Council of Norway through its Centres of Excellence funding scheme, project number 223268/F50.

References

1. Gjelsvik R (2014) Duration and long-term management of ^{137}Cs in livestock products in Norway. In: Proceedings of the international conference on radioecology and environmental radioactivity, Barcelona, Spain, 8–12 September 2014
2. Liland A, Skuterud L (2013) Lessons learned from the Chernobyl accident in Norway. In: Oughton D, Hansson SO (eds) Social and ethical aspects of radiation risk management. Elsevier Science, Amsterdam, pp 157–176. ISBN: 978-0-08-045015-5, ISSN: 1569–4860
3. Liland A, Lochard J, Skuterud L (2009) How long is long term? Reflections based on over 20 years of post-Chernobyl management in Norway. J Environ Radioact 100:581–584. doi:10.1016/j.jenvrad.2009.04.006
4. NOU (2012) Norges offentlige utredninger (2012:14) Rapport fra 22. juli-kommisjonen [Report from the Commission on 22nd July]. Oslo, 2012 (in Norwegian)
5. NRPA (2007) Bulletin 8•07. The Kyshtym accident, 29th September 1957. Norwegian Radiation Protection Authority, 2007. ISSN 0806-895x
6. Skuterud L, Thørring H (2012) Averted doses to Norwegian Sámi reindeer herders after the Chernobyl accident. Health Phys 102(2):208–216

Chapter 20
Another Cost of 9/11

Carol Kessler

Abstract The 9/11/2001 terrorist attacks have long-standing psycho-social impacts that stay with the American public today. These impacts derive from the fear, insecurity, vulnerability, and incomprehensibility that resulted from the shock of 9/11. These impacts cannot be forgotten and create a constant anxiety of the terrorists next strike. These residual fears make us more vulnerable to panic with each succeeding terrorism event. These impacts were and are reinforced by the way the USG has responded to 9/11 - with fear itself. It has invested in two wars including a huge and continuing global war on terror and overt, massive security measures on U.S. soil to protect Americans from nuclear, radiological, biological or chemical weapons use by a terrorist. These fears have hindered Americans' ability to recognize that such attacks, while having disastrous effects, are low probability events due to the actions of the USG and other partner governments to prevent such attacks. These reactions in Americans may be due to their relative unfamiliarity with what the USG is achieving in terms of improved security in the U.S. It is important for Americans to understand what security improvements are in place so they can find the collective strength to reduce their anxiety and handle future events with resilience. Americans also need to become capable of judging when these security investments should be decreased because they are no longer cost-effective. There are many other investments desperately needed to improve the overall health and well-being of Americans. The recent overreaction by the U.S. public and government to the introduction of the Ebola virus on the U.S. mainland is given as an example of the lasting impact of 9/11 psycho-social impacts. The overreaction was caused by unwarranted fears that led the public and some government leaders to dismiss the science and common sense and take needless precautions which reinforced Americans fears rather than dissipating them.

The impacts of 9/11/2001 terrorism events have been discussed in many fora. There were short term and there are long term impacts: economic, technological, political

C. Kessler (✉)
NorthRaven Consulting, LLC, Baiting Hollow, NY, USA
e-mail: cakessler@aol.com

S. Apikyan, D. Diamond (eds.), *Nuclear Terrorism and National Preparedness*,
NATO Science for Peace and Security Series B: Physics and Biophysics,
DOI 10.1007/978-94-017-9891-4_20

and psychological/social impacts. I have been thinking about these from my own perspective and how I have seen my life changed as a result of 9/11. It is easy to note some of the impacts. Thirteen years ago, America wasn't engaged in any foreign wars. We deported half as many immigrants as we do today. And getting through airport security was a total breeze.[1] Now getting on the plane to come here was only bearable because I now have "TSA precheck" which exempts me from the longest lines and most intrusive inspections.

Some of the most direct impacts of 9/11 hit us economically. These included the lives lost, billions of dollars in damage, stock market crash, and also the costs of creating whole new security structures in government and private industry to shore up the mainland United States (U.S.) against terrorism threats. New technologies were developed for surveillance, intelligence collection, security in airports, radiation detection in airports and on borders, even bridges, such as in New York City, and no doubt elsewhere around the U.S. These new technologies were important but costly.

I have become concerned about some worrisome impacts; some of the psycho-social impacts whose effects are significant, sustained and damaging today and to the long term interests of the United States. These are the feelings that still come to mind when Americans think of 9/11 and color our views of our world today: Feelings such as fear, uncertainty, vulnerability; reactions that the events were somehow, incomprehensible, unforeseeable, that we were unprepared and perhaps can never be well-prepared makes innocent people fear that the worst will happen, and believe that no one can be trusted anymore. These are not the reactions that contribute to rebuilding a nation after a devastating event and they remain as a legacy of 9/11 in our collective psyche in the United States.

When the USG commenced its response to 9/11, it created the global war on terror to counter the unpredictable and highly dangerous world of terrorism. Legislation was enacted that progressively granted more and more rights to the government to help protect Americans. The legislation also affected our right to privacy at home, leaving some individuals feeling that their civil liberties and rights were/are being infringed upon. While, others, large numbers of Americans, felt that they would only be safe with strong and intrusive government measures due to their deep sense of vulnerability. As we know, a new government department, the Department of Homeland Security, was established to design and build extensive new security arrangements to protect against homeland terrorist attacks. 9/11 also raised the possibility again that terrorists might use a weapon of mass destruction in their attack and so the Department was tasked, if at all possible, to prevent them.

Internationally, the fears from 9/11 drove the USG to do more than it had previously felt necessary to prosecute a war. The Global War on Terrorism actions, including waterboarding prisoners of war, went beyond legal boundaries we had previously set for ourselves and the urgency and impunity with which the USG pursued these actions, arguably contributed further to the insecurity felt by many in

[1] http://Blogs.kqed.org/lowdown/2012/09/11/911-turns-11-three-major-lasting-changes

the U.S. public. Seeing the government act as desperately as it did, sustained the fear and, in turn led many Americans to accept whatever measures the USG proposed to protect them with little thought to their potential or real costs. For most Americans, this is a pretty high bar to have jumped just a few years after 9/11, and yet we did – out of fear.

This fear, vulnerability, and insecurity still weigh heavily on the American psyche today which makes it hard for Americans to handle any new acts of violence as these are reminders of our loss of safe and secure everyday lives.[2] And what are other consequences from such fear? As David Brooks put it; "People seek to build walls, to pull in the circle of trust... Fear, of course, breeds fear. Fear is a fog that alters perception and clouds thought. Fear is, in the novelist Yann Martel's words, "a wordless darkness.""[3] America's fear has fed on itself as new horrifying events have taken place since 9/11. These are compounding our insecurity, our sense that our world is out of control and our inability to "cure" ourselves and focus on things that can rebuild our national psyche in positive ways. Americans have lost the ability to gauge the relative dangers these events bring to our lives. We need the government's help to rebuild our perspective and surmount the insecurity that interrupts our ability to balance our hopes and fears. We need to regain its assurance that Americans can manage its future security successfully.

At the time of 9/11, Americans had many choices on how to respond and pull ourselves back together. Our government chose a hard course, to go it alone. America has paid a high price for this decision. As of 2011, it was estimated that the cost of involvement in the war in Afghanistan was roughly $440 billion, while the Iraq war had come to an even larger sum of nearly $790 billion. During this same time period, the Homeland Security Department's budget changed from $16B to $43B for a total of nearly $649B from 2001 to 2011.[4] These expenditures make the 9/11 global war on terrorism one of the most expensive wars in our history. Especially because on top of the budgetary costs, our allies began to call into question how the U.S. was using its power and our reliability as their partner. They viewed U.S. leadership as too reactive, lacking confidence, and at times quite erratic. The USG rejected the help of most of our allies at the time of 9/11 and acted as if we could no longer trust them. We decided we could not afford to cooperate; we needed instead to strike back through an offensive foreign and military policy. For many, the U.S. lost its widely accepted role as the "leader of the free world", because we chose to go it alone.

---

[2]"We keep getting reminders of how American interests are under attack. Which reminds us of our vulnerability. The sense of security we had up until 9/10, that nobody could touch us here in America, is gone." noted Prashant Gajwani, M.D., associate professor of psychiatry, University of Texas Health Science Center at Houston in his paper on the permanent scars on the American psyche from 9/11; http://consumer.healthday.com/mental-health-information-25/psychology-and-mental-health-news-566/9-11-left-permanent-scars-on-the-american-psyche-656551.html

[3]http://www.nytimes.com/2014/10/21/opinion/david-brooks-what-the-ebola-crisis-reveals-about-culture.html?emc=eta1&_r=0

[4]http://costsofwar.org/article/homeland-security-budget

As we approach the end of 2014, it is easy to see the impact of these 9/11 impacts on the U.S. public today. An example is the U.S.'s reaction to the Ebola virus attacks in West Africa. We are reacting with fear; we feel so vulnerable that we are allowing fear to override the science and our common sense. Americans are allowing the government to take actions that are unbalanced and cause their fear to deepen instead of dissipate. As David Brooks commented shortly after the Ebola reaction in the U.S. careened out of control, "There were the hundreds of parents in Mississippi who pulled their kids from school because the principal had traveled to Zambia, a country in southern Africa untouched by the Ebola outbreak in the western region of the continent. There was the school district in Ohio that closed a middle school and an elementary school because an employee might have flown on the same plane (not even the same flight) as an Ebola-infected health care worker."[5] Some Americans worried, and maybe some still do worry, that Ebola will inevitably hit the U.S. mainland as an epidemic no matter what we do. And government officials, sensing the depth of public fear, were taking extreme measures in their zeal to protect the public. Instead of working in solidarity with the federal officials, Governors took their own actions. The New Jersey and Maine governors told a young nurse, whose training and awareness were as high as anyone's about the Ebola risks, that she could not be trusted to take care of herself. New Jersey quarantined her for days and when she was released to go home,[6] Maine told her to stay home for 21 days even though she had no symptoms.[7] The Governor of Louisiana ordered the organizers of two major international scientific meetings of medical researchers in New Orleans, scheduled for the week of November 7, 2014, to bar scientists from the World Health Organization and the Centers for Disease Control and Prevention (CDC) who had treated Ebola patients or been in Guinea, Sierra Leone, or Liberia from participating in the meetings. Meanwhile, two infectious disease meetings in Europe went ahead as planned.[8]

What is the general public led to think as a result? Perhaps the U.S. was not prepared to deal with Ebola; it could not protect its citizens against this disease, and that possibly no one knows how to? This uncertainty increased the public's insecurity and pervaded their thinking hastening their loss of perspective on what the real risk was. Many Americans did not take comfort, much less hear, the head of the CDC say every night on television, "The bottom line with Ebola is we know how to stop it: traditional public health. Find patients, isolate and care for them; find their contacts; educate people; and strictly follow infection control in hospitals.

[5]http://www.nytimes.com/2014/10/21/opinion/david-brooks-what-the-ebola-crisis-reveals-about-culture.html?emc=eta1&_r=1

[6]http://www.nbcnews.com/storyline/ebola-virus-outbreak/new-jersey-releases-nurse-quarantined-suspected-ebola-n234661

[7]http://www.nbcnews.com/storyline/ebola-virus-outbreak/kaci-hickox-humbled-maine-judges-ebola-ruling-n238326

[8]Science, 7 November 14, vol 346, Issue 6210, P. 680.

Do those things with meticulous care and Ebola goes away."[9] Instead, parents let schools scare their children by sending them home or keeping them out, home themselves. The public supported their governors' calls for quarantining people getting off planes from West Africa. Americans convinced ourselves that only taking drastic measures domestically could stop the Ebola onslaught.

Our barometer for managing uncertainty and risk appears to be off kilter. It appears that the psychological and social impacts of the 9/11 attack are being intensified every time Americans encounter another act of terror – when Virginia Tech went on lockdown, when a midnight showing of Batman in Colorado turned deadly, when a gunman entered an elementary school in Connecticut, and when explosions shattered the finish line of the Boston marathon. Strikingly, "nearly three out of four Americans say that terrorism prevention is equal to or more important a priority than things like the preservation of families, immigration, healthcare, unemployment and education. Even [12] years after the 9/11 attacks, it would seem the threat of terrorism remains a powerful public motivator in America."[10]

Is this the country we wish to remain in? Must we continue to harbor the anxiety of 9/11, the sense of foreboding that some horrific terrorism event is still imminent, that the USG does not understand what scale of threat to prepare for, and thus does not have the tools to deal with them. The 9/11 terrorist attacks ignited fears in the technical community that terrorists might turn to weapons of mass destruction (WMD): either nuclear, radiological, biological or chemical. The U.S. government invested in how it might thwart such WMD attacks, and possibly prevent them, rather than mitigate the effects after their use. And as noted above, we have spent billions of dollars since analyzing the possibilities, the probabilities and building up our technical abilities to respond to a WMD attack through the Department of Homeland Security.

Since 9/11, the American technical community, and hence the public, has believed that terrorism, including nuclear terrorism, is a more probable event. Do we know what the real probability is of such an event? Are the consequences of such an act so high that it warrants the continued investment? How much protection is enough? I don't have the answer, but I have enough knowledge of what has been done and is being done by the U.S. government to prevent or mitigate such attacks. We have some of the answers and it may be time to reassess. If nothing else we owe it to the American public to let them know how much protection has been developed so that this knowledge can be used to help rebuild our perspective on threats to our security.

After a decade of investment, it is time to communicate more effectively with the public about the progress we have made in preparing to respond. It is time to reassess whether we can afford to ramp down our programs and concentrate these

[9] http://www.cdc.gov/media/releases/2014/p0806-ebola.html

[10] https://www.barna.org/barna-update/culture/626-the-emotional-and-spiritual-aftermath-of-9-11#.VF5pVVQo7DA

massive resources on other means to strengthen the United States. We need the U.S. government to evaluate what America could do to instill a policy and practice that enables Americans to live more balanced lives. Importantly, Americans also need to gain the resilience to maintain that balance in the face of the types of terrorism we are most likely to face. That could instill the sense that we can return to a more secure and balanced life.

It could be difficult to find the words needed to temper the U.S. public's fears because the nuclear terrorism threat is a low probability event. Could they take comfort in what the USG investment has enabled in terms of protecting us in the event of an attack? When do the public know the investment is enough? What does the U.S. response to Ebola and the fears it engendered tell us? Can the level of fear among Americans be checked by educating the public about our preparedness and can we find words to explain the confidence they can have in the current level of preparedness. It remains important for the government to remain vigilant and continue to invest in our preparedness, but is it necessary for the public to remain fearful in their daily life? Wouldn't the United States benefit if Americans could believe that we are better prepared to protect ourselves from many terrorist events.

The American public, once educated on the U.S. strengths in science and technology, can rely on this knowledge to reject sensationalist news such as the Ebola panic. The fear and anxiety such panic generates saps our national energy and resources and leave us less able to best protect ourselves from real threats. Unfortunately, this won't be easy. "Skepticism about science and expertise and authority has a pretty big constituency out there," said Ross K. Baker, a professor of political science at Rutgers University. It is not enough for policy makers to be right on the science, he said; they must also find a way to reassure "people who are all too ready to interpret expert opinion as elitist and condescending."[11]

So what can we do? A concentrated effort is needed by the science and technology community to educate the public so they understand how much has been done since 9/11, the successes we have had in stopping terrorist attacks and in preparing for their possibility. The U.S. government needs to emphasize the importance of working on problems to solve them rather than letting them overwhelm us, and most important, educate the public on the relative risks of terrorism versus other threats in their lives. U.S. government needs to take credit publicly for technology that has improved our border security, our transportation networks, and our response capabilities if an attack gets through. The government should explain how these achievements address the weaknesses exposed by 9/11 and later events, and in language the public can understand, yes, no acronyms. Those of us with expertise and experience need to call out the press rather than support it when it seeks to sensationalize an event. The press should give the facts and help the public to appreciate the complexity involved, and with that knowledge be able to understand why the USG does know how to better protect them. It is critically

[11]http://www.nytimes.com/2014/11/01/us/alarmed-by-ebola-public-isnt-calmed-by-experts-say-.html

important that the media help to rebuild public trust, but, no doubt, it will take a great deal of cultural change to get the media away from the desire to sell sensationalism.

And what has been done? The USG has thwarted many attempted terrorism attacks since 9/11. It has developed better tools: better intelligence methods, better security practices, better security technology. Technology, in fact, has advanced significantly since 2001. New technology has made better security possible through better communications technology, better satellites, better radiation detectors, and better computers just to mention a few. The U.S. has and is practicing emergency response procedures at the local, state and federal level. It holds field exercises to test the joint response capabilities. It has made agreements with countries all over the world on emergency response and on use of improved security technology and security practices. The U.S. first responders protect every major event from Super Bowls to meetings of the UN General Assembly from possible terrorist attacks. The USG has sophisticated network analysis capabilities that help track terrorists' movements and hence it has been able to stop some terrorists in their tracks. Importantly, these successes need to be emphasized by responsible politicians who will reinforce the science and technology community messages. We need this education campaign to begin in schools to help students and their parents participate in the national dialogue.

We need another paradigm for managing the United States in the twenty-first century than fear, insecurity and vulnerability. We need to determine whether terrorism really needs to remain our first priority. We can't forgo important measured expenditures, but how much is really needed? Is Homeland Security investment worth losing our reliable network of highways, our science and technology leadership, our ability to feed ourselves and others. The tradeoffs are constantly debated, but with little information on our real capabilities in 2014 to protect ourselves from WMD terrorism. We can change that. Those of us with the knowledge must make it habit to speak openly and regularly with the public to explain what we have done and are doing to make ourselves more secure.

Chapter 21
Assessment of Dose and Risk for Public from Potential Exposure Using WinMACCS and JRODOS Codes

Alla Dvorzhak, Juan C. Mora, and Beatriz Robles

Abstract Potential exposure is prospectively considered exposure that is not expected to be delivered with certainty but that may result from an anticipated operational occurrence or accidents because of equipment failures, operating errors or external initiators. Potential exposure situations are events of probabilistic nature and the possible radiological impact as a comprehensive view shall be considered. In this paper a Probabilistic Safety Assessment (PSA) Level-3 for potential exposure was fulfilled using computer code MACCS (MELCOR Accident Consequence Code Systems). Additionally the deterministic modeling of consequence analysis for the critical meteorological conditions was fulfilled by the JRODOS decision support system (Real-time On-line Decision Support system for off-site emergency management in Europe). The framework for doses and risk assessment from potential exposure of accident releases are presented. Two approaches of safety acceptability analysis are demonstrated: the estimated doses dependent on distances of a release and other one is risk estimation. Both approaches are complementary, although the risk approach takes into account more aspects. So, the usage both of them can be considered an advantage. Comparison of the consequences with the risk curve acceptability criteria is shown.

21.1 Introduction

The potential exposure is prospectively considered exposure that is not expected to be delivered with certainty but that may result from an anticipated operational occurrence or accident at a source or owing to an event or sequence of events of a probabilistic nature, e.g. equipment failures, operating errors or natural phenomena (such as hurricanes, earthquakes and floods) [1].

A. Dvorzhak (✉) • J.C. Mora • B. Robles
Radiation Protection for Public and Environment, CIEMAT, Madrid, Spain
e-mail: alla.dvorzhak@ciemat.es

S. Apikyan, D. Diamond (eds.), *Nuclear Terrorism and National Preparedness*, NATO Science for Peace and Security Series B: Physics and Biophysics, DOI 10.1007/978-94-017-9891-4_21

To provide a framework of common understanding for potential risk assessment and prepare a guidance to assess the risk of accidental releases, to be used when comparing different alternatives of Innovative Nuclear Systems (INS) this study was made in the framework of the IAEA INPRO Collaborative Project [2].

21.2 Probabilistic Safety Assessment (PSA) Level-3

The developed methodology is based on the PSA Level-3: assessment of the off-site consequences and estimation of the risks to the public [3]. The phenomena such as atmospheric dispersion, deposition of airborne materials, resuspension, migration through the food chains and other have been considered in the appropriate codes.

For the probabilistic consequence analysis and modeling of the mentioned processes the WinMACCS code version 3.7.0 has been used [4], additionally the JRODOS code has been applied for the deterministic modeling of consequence analysis for the critical meteorological conditions.

WinMACCS code, designed primarily as a probabilistic risk assessment (PRA) tool, accounts for the uncertainty in weather, and so random weather sampling addressed the uncertainty in health effects from accidental releases caused by weather variability. At the same time the code permits evaluate the impact of uncertainty of the model parameters by introducing random sampling distribution for key model parameters [4]. On Fig. 21.1 the main concept of the modeling with WinMACCS code is presented.

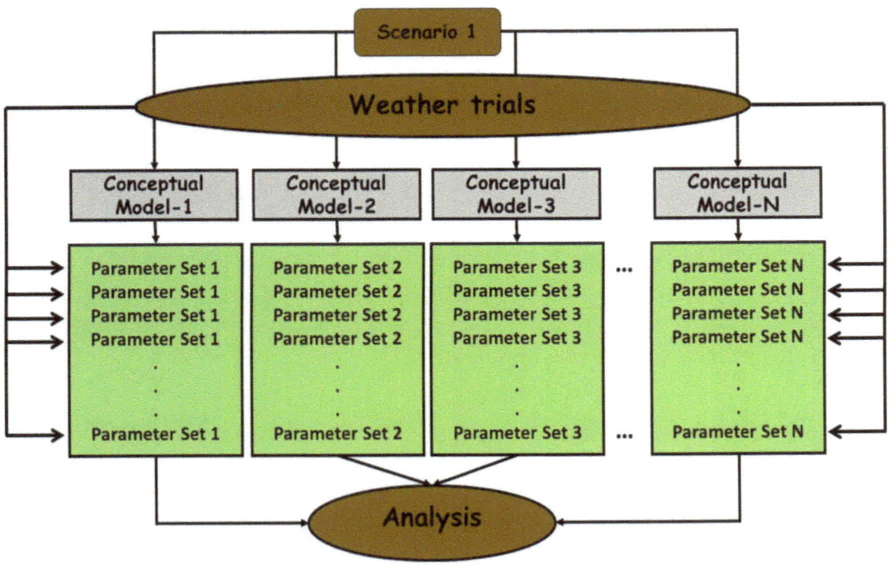

Fig. 21.1 Conceptual schema of the modeling with WinMACCS code

In the beginning of the study an appropriate scenario was chosen with the meteorological data accumulated for at least 1 year with hourly step interval, the corresponding source term, and other input information and model parameters. Based on this information the iterations for considered weather sampling (8,760 sequences) are fulfilled with calculations for each trial of the atmospheric dispersion and deposition; estimation of the doses, health effects. This modeling takes into account the weather uncertainty in correspondent location and in such way the probability of the consequences is estimated. Thereafter the received array of results is subjected to the statistical reprocessing. If uncertainty of the model parameters is taken into account, more cycles of iterated calculations should be made with the same procedure of the consequences estimation.

The above described consequences are evaluated on a grid (r, θ) around the release location. The results are produced on each grid elements for large amount of weather conditions. This produces a distribution of the individual risk at each grid element. For each distribution a mean value of individual risk can be obtained for each grid element. But it has become standard practice to further average them over all directions of the wind rose, rather than presenting these mean values of the individual risk for each grid element. Some studies are based on the approach of maximum doses on fixed distances from the point of release [5, 6]. This study is also centered on the peak doses over all directions of wind rose at each distance and each weather sequences (see Fig. 21.2).

Peak dose means the maximum dose around the compass under all meteorological sampling options (see the formula (21.1)).

$$D_{ri}^{MAX} = MAX_{j=1}^{K} D_{rij} \tag{21.1}$$

Being:

D_{ri}^{MAX} dose over all directions of the wind rose on the distance r from point of release for weather sequence i;

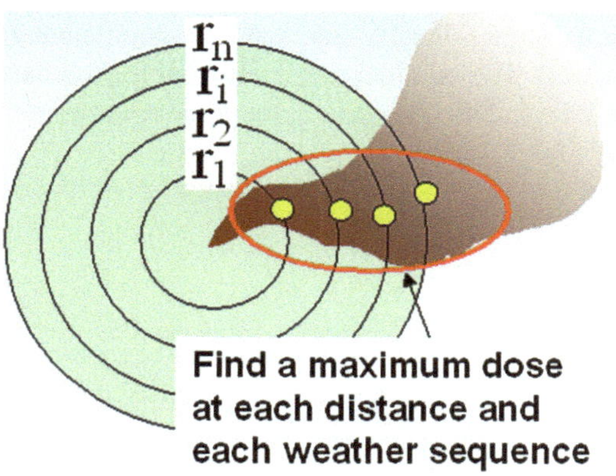

Fig. 21.2 Peak values over all sectors of the ring at certain distances from release [5]

D_{rij} – Dose value at distance r from point of release, at sector j, for weather condition *i;*
r – Index of grid element by the distances from the point of release;
i – Index of weather sequences;
j – Index of grid element by the sectors at certain distance;
K – Amount of the sectors (in the one ring);

Then the samples of the D_{ri}^{MAX} values are undergone to statistical reprocessing such as estimation of the percentiles, complementary cumulative distribution functions (CCDFs). The CCDF consist of plots on log-log graphs of exceeded probability versus consequences.

21.3 Risk Indicators

One way to consider potential exposures is to use a measure of risk, i.e to estimate the percentiles of doses to evaluate the risk, which estimates the probability that a certain individual is accidentally exposed to ionizing radiations due to a certain category of accident at a plant and results in health effects. As general view the risk can be expressed by such self-explained formula:

$$Risk = Threat\ Rating \times Consequences\ Values \times Vulnerability\ Rating \qquad (21.2)$$

The 'Threat Rating' in our case means probability of the accident occurrence, 'Consequences values' means the doses estimated, 'Vulnerability Rating' is the sensitivity or weakness of the site of plant, which is taken into account through the weather variability, and sensitivity of the adult person to intensity of exposure.

So, the risk from the release occurred at a specific location (with typical meteorological conditions in this location) was calculated as product of 95th percentile of the assessed dose (which itself includes risk concept, as was received based on the weather sampling), the probability of occurrence of the release of considered category and the risk coefficient for stochastic effects (see formula (21.3)). The nominal risk coefficient for stochastic effects was taken as 0.057 Sv^{-1} [7]. Probability of occurrence of the release category was considered 1.5×10^{-6} [8].

$$Risk = P(RC)^n \cdot D \cdot f(D) \qquad (21.3)$$

Being:

Risk – Individual risk;
$P(RC)^n$ – Probability of occurrence of the Release category n;
D – Effective dose (95th percentile based on weather sampling);
f(D) – Nominal risk coefficient for stochastic effects

The results of the dose and risk estimation are presented in the following sections in the exercise description.

21.4 Scenario and Results

The source term of the accidental release was taken from the SOARCA project for Surry NPP [8] which is assumed as postulated radionuclide release to the atmosphere. The sequences initiated by external event as seismic incident was considered. The short-term station blackout (STSBO) with core damage frequency (CDF) to 1.5×10^{-6} per reactor-year (pry) initiated by an earthquake (0.5–1.0 g peak ground acceleration (pga)) was assumed for analysis of the environmental impact and doses consequences. This event results in the loss of offsite power and failure of onsite emergency alternating current (AC) and direct current (DC) power, resulting in an station blackout (SBO) event where neither onsite nor offsite DC and AC power are recoverable. All systems dependent on AC power are unavailable, including the containment systems (i.e., containment spray and fan coolers).

Meteorological data were derived from one Spanish meteo station, which describes the weather condition during 1 year with 1 h interval measurements of the wind direction, wind speed, atmospheric stability, and precipitation rate.

21.4.1 Results of Doses Estimation

The doses through several pathways including external from cloudshine, groundshine, inhalation, immersion and deposition onto the skin were estimated. Separately the ingestion pathway from contaminated food was analyzed. One of the endpoints in this study is the total effective doses (ICRP60) based on the effective dose coefficient from [9]. The age category 'adults' is taken into account and the effective dose is calculated for an integration period of 50 years for inhalation, groundshine and ingestion pathways, and during plume passage for short-term pathways. As it was mentioned before the 'peak doses' were selected and analyzed.

In accordance to TECDOC-1575, vol. 8 (2008) [10], "Safety of Reactor", UR1.5 "A major release of radioactivity from an installation of an Innovative Nuclear System (INS) should be prevented for all practical purposes, so that INS installations would not need relocation or evacuation measures outside the plant site, apart from those generic emergency measures for any industrial facility used for similar purpose." So, following investigation is based on the analysis of these criteria as need of relocation and evacuation countermeasures. The lifetime overall doses and doses integrated for 7 days were analyzed. The ingestion doses are not presented here and were calculated apart for possibility to consider scenario that uncontaminated food and water can be supplied and that the public would not eat radioactively contaminated food. In accordance to Spanish legislation evacuation countermeasure is based on the criteria 50 mSv of 7 days integrated dose [11], so estimation of such doses was fulfilled to check if evacuation is necessary (see Table 21.1).

Table 21.1 Peak 7 days' integrated doses on spatial grid (Sv)

Distances from the release point, km	Percentiles of doses (Sv)						
	Mean	90th	95th	99th	99.5th	Peak conseq	Peak Trial
0–0.1	**1.99E+00**	**2.36E+00**	**2.53E+00**	**2.99E+00**	**3.08E+00**	**3.61E+00**	6679
0.1–0.5	**5.60E−01**	**7.29E−01**	**7.68E−01**	**8.66E−01**	**9.13E−01**	**1.08E+00**	6417
0.5–1.0	**2.68E−01**	**3.48E−01**	**3.78E−01**	**4.57E−01**	**4.96E−01**	**5.90E−01**	6417
1.0–1.5	**1.86E−01**	**2.66E−01**	**3.02E−01**	**3.36E−01**	**3.52E−01**	**4.53E−01**	6335
1.5–2.0	**1.51E−01**	**2.30E−01**	**2.60E−01**	**3.18E−01**	**3.34E−01**	**4.42E−01**	6333
2.0–3.0	**1.16E−01**	**1.80E−01**	**2.17E−01**	**3.01E−01**	**3.17E−01**	**4.19E−01**	7818
3.0–4.0	**8.62E−02**	**1.40E−01**	**1.70E−01**	**2.38E−01**	**2.68E−01**	**3.50E−01**	6332
4.0–5.0	**6.69E−02**	**1.14E−01**	**1.35E−01**	**2.00E−01**	**2.13E−01**	**3.00E−01**	7817
5.0–6.0	**5.33E−02**	**9.46E−02**	**1.11E−01**	**1.55E−01**	**1.78E−01**	**2.46E−01**	7817
6.0–8.0	3.92E–02	**7.10E−02**	**8.64E−02**	**1.12E−01**	**1.21E−01**	**1.80E−01**	7817
8.0–10.0	2.77E–02	**5.07E−02**	**6.14E−02**	**8.57E−02**	**9.70E−02**	**1.27E−01**	7817
10.0–16.0	1.56E–02	2.79E–02	3.39E–02	**5.04E−02**	**5.25E−02**	**6.56E−02**	7817
16.0–20.0	9.29E–03	1.61E–02	2.04E–02	3.02E–02	3.16E–02	4.04E–02	6350

With *red color* the values greater than 50 mSv (limit for evacuation) are marked

It can be seen that for this scenario, evacuation could be necessary till the distance 10–16 km.

Other method to check the doses criteria for evacuation is shown at the following graph of dose percentiles together with dose limit curve (Fig. 21.3). Such family curves can help in understanding the results and facilitate in the making decisions. There is shown for percentile 99th the distance where the dose exceeds the limit is 16 km, what means that only for 1 % of the cases this distance can be exceeded. This curve is very close to 99.5th percentile and the curve of max consequences, so is negligible probability that the criterion is exceeded the distance more than 16 km. On the other hand for 50th percentile the distance of exceeding the dose criterion is 6 km, which is high frequency and large distance, so evacuation on off-site area, out of a fence of a site will be necessary. So, in accordance to this scenario INS should be not acceptable.

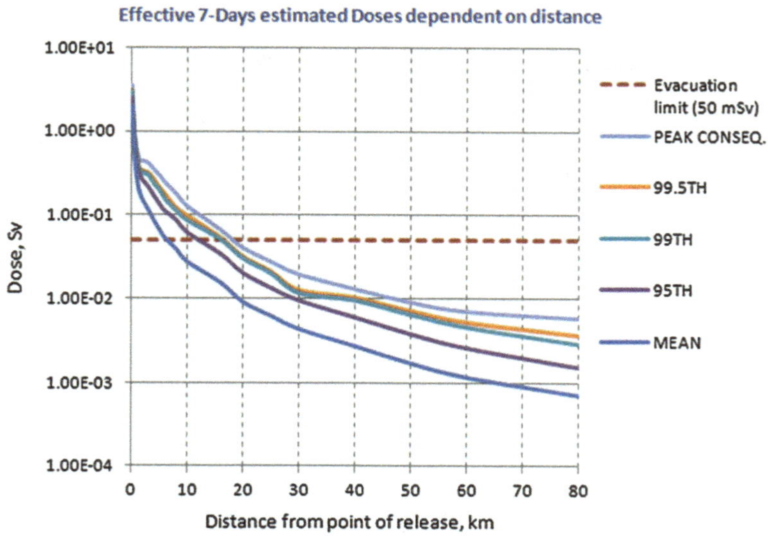

Fig. 21.3 Effective 7-Days estimated doses dependent on distance from point of release

This approach is similar to that presented in [12], but with different objective of the assessment of potential exposure. In this study the attempt was made to find out if the design was appropriate and if installation was acceptable from point of view of a public and environment risk, while in [12] the main objective was the definition of Emergency Planning Zones or revising of emergency management requirements for new generation reactors.

Other useful information can be extracted from the results of the 'Peak trial' (Table 21.1). These values inform about the trial of weather conditions, which can give the highest consequences. As we made simulation for a whole year, the total set of 8760 trials is analyzed. It can be seen that on the different distances from the point of release 7-days' doses achieved maximum values for 6332–8512 trials, meaning that the critical period is the autumn time (from September to December). For lifetime dose (except ingestion pathway) more critical time of the release is considered from April till September.

Most critical trial for 7-days doses for the distances 4–16 km from point of release (Table 21.1) is the trial 7817, which can lead to the need for extensive evacuation. This trial was analyzed in more detail with JRODOS system [13], as it permit the use of more sophisticated atmospheric dispersion modeling, plainer visualization using standards ESRI of geographical maps etc. We demonstrate this approach as example of graded dose and risk consequence analysis using two tools WinMACCS and JRODOS (one of the examples of JRODOS using for the analysis of the accident at Fukushima Daiichi NPP can be seen in [14]).

The data base of RODOS includes the inventories of all European NPPs, Surry NPP is not available, and therefore the NPP was included into the RODOS database.

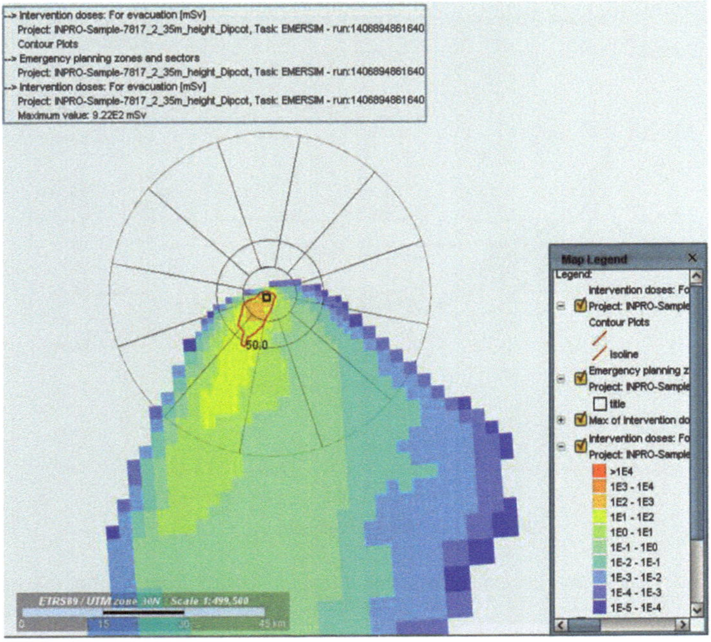

Fig. 21.4 The effective doses integrated for 7-days doses after accident release and isoline of 50 mSv (Evacuation zone) calculated by Dipcot of JRODOS system

The calculation was made for a fictitious site with defined source term which was used as an input the fractions released of inventory for selected release group of MELCOR_10_GROUP. As the land use type the 'everywhere grassland or unidentified area' were used. Actual geographical and land use GIS maps of the site should be used for real case study calculation. On the Fig. 21.4 the results of 7-days effective dose for this scenario simulated by JRODOS system and DIPCOT atmospheric dispersion model [15] are presented. Red line marks the zone where the dose exceeds 50 mSv and where evacuation could be necessary. This zone is extended to 10 km. Figure 21.5 presents the evacuation zone by RIMPUFF code of JRODOS [16] where evacuation zone can be located in the range from 10 to 20 km approximately. It should be taken in mind that this evacuation zones were derived for the more critical weather trial and that conservative modeling conditions were considered in the modeling with JRODOS system. This exercise was curried out to demonstrate the possibility of using other tools with more detailed modeling of certain atmospheric conditions, while WinMACCS code was used for probabilistic modeling.

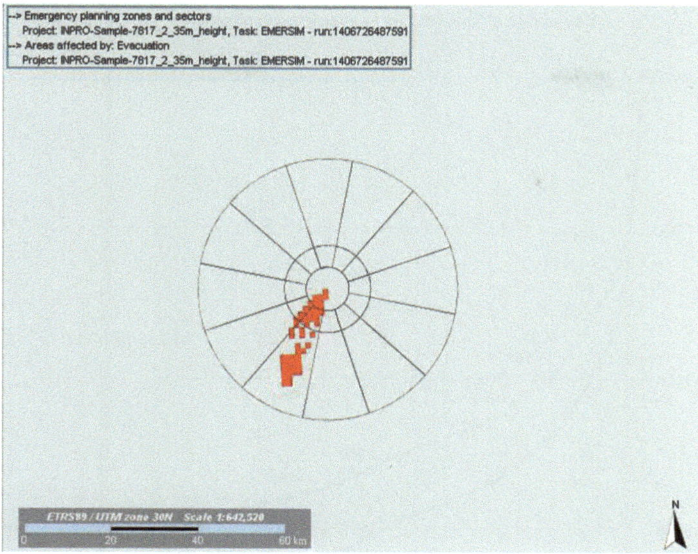

Fig. 21.5 Evacuation area derived from model Rimpuff of JRODOS system. Represented zones: 5, 10, 30 km. Evacuation extension till 24 km

21.4.2 Risk Estimation

The risk was calculated as product of lifetime effective dose (95th percentile), the nominal risk coefficient for stochastic effects, and the probability of occurrence of the release category (see formula (21.3)). The results of the estimated risk are presented on the Fig. 21.6. The risk curve constructed for 50th percentile of doses is also presented for comparison. So, in this graph the risk, as function of distance from point of release, is shown.

Additionally the values of the doses (95th percentile) were used on the curve acceptance criteria developed by Argentina [17, 18] with changes for adapting it to Spanish legislation (Fig. 21.7). The values of the doses for different distances from the point of release are shown on graph and it can be seen, the more far we are from point of release, more far we are situated from the risk curve on the acceptable area and respectively more safely. Only one point falls directly into non acceptable area (0.5 km) and other one (1 km) falls on the limit risk line. Therefore it can be estimated that risk value is exceeded for very short distances to the point of release (which can be practically without habitants and can be considered exclusion zone around of the plant). So, the nuclear installation could be acceptable in such assumptions.

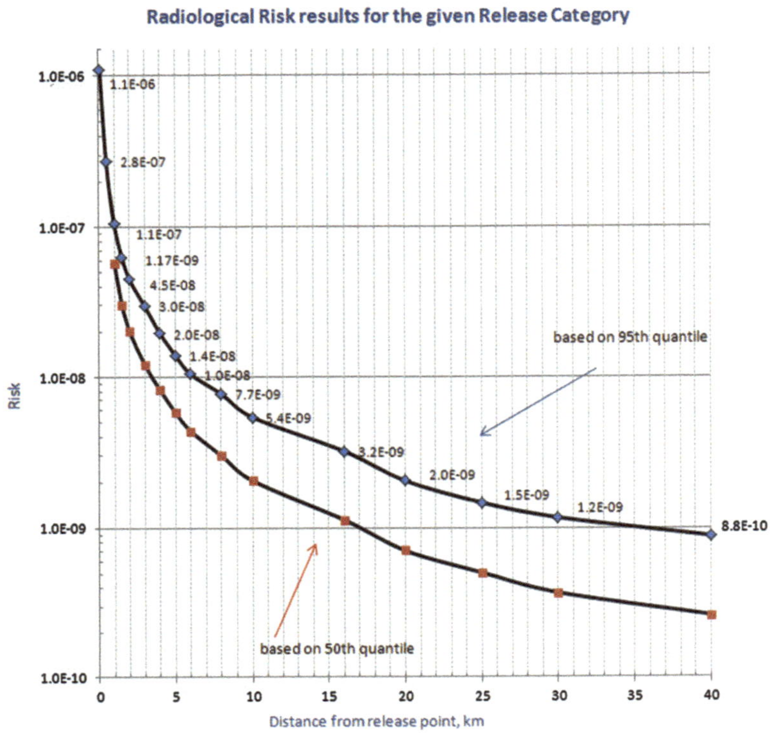

Fig. 21.6 Risk curve based on the 95th and 50th lifetime dose percentiles

21.5 Discussion

This methodology can be proposed as a risk-informed approach to analyze the acceptability of a new installation.

First step is the selection of a scenario for certain release category with its frequency. In practice analysis should be made for the whole set of release categories, but in this work was used as example only one selected category (see section of source term description). Appropriate source term for this category should be defined. Also typical year meteorological data must be obtained.

Second step is modeling (as PSA Level 3) the doses curves family (doses vs. distance) as percentiles based on the sample of the weather conditions, 50th, 95th, 99th and peak consequences.

Third step is identification of the distances where doses exceed the criteria for countermeasures as evacuation or relocation.

The fourth step is analysis and solution. If the doses do not reach the criteria at any area, than installation is acceptable. If the doses exceed the criteria, the distances should be analyzed. If dose criteria are exceeded in off-site zone, it means that evacuation or relocation of the public will be necessary, so the installation would be not acceptable.

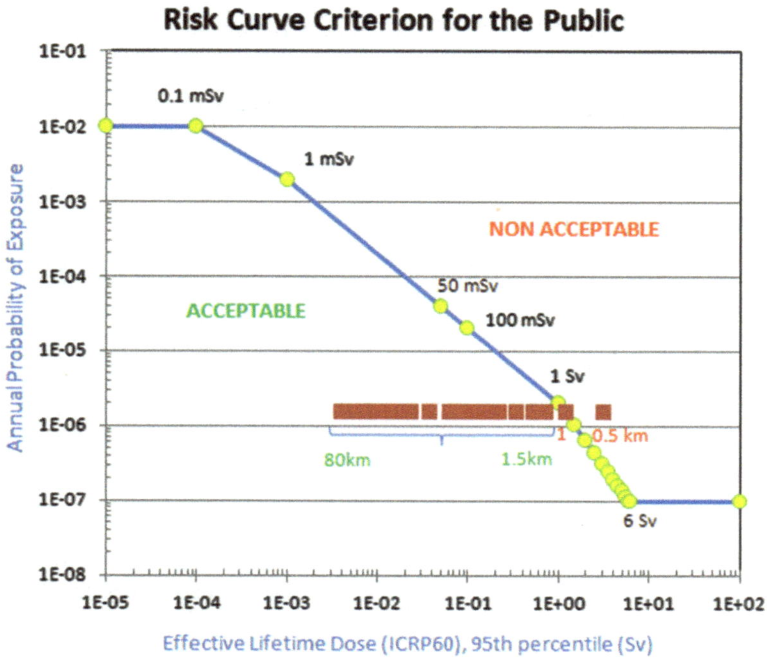

Fig. 21.7 Comparison lifetime doses depending on distance from point of release with risk curve criteria as function from annual probability of the release

Also a risk can be calculated and compared with the risk curve acceptability criteria, which consider in coordinate system of the effective doses and the annual probability of accidental sequences [18]. The two approaches for the acceptance criteria for Innovative Nuclear System (dose and risk) complement each other. Whereas the dose approach is more clear and understandable for experts in radiological protection, the risk approach takes into account more aspects. So, the usage of both approaches can be an advantage. In this study it seems that doses criteria are a little bit more restrictive than risk criteria.

21.6 Summary

Two state of the art tools have been used in the study with different focus of the application: WinMACCS code with probabilistic approach, with broader view on the problems, including probabilistic aspects, but simpler models, and JRODOS code with emphasis on a concrete site, with standard ESRI geographical maps as background for results and for input information. The risk-informed approach for acceptability of INS was analyzed on a predefined scenario. A family of curves of doses for analysis of countermeasures was performed.

References

1. IAEA (2014) Radiation protection and safety of radiation sources: international basic safety standards general safety requirements. IAEA Safety Standards series no. GSR Part 3. IAEA, Vienna
2. International Atomic Energy Agency (2014) "Environmental impact of potential accidental releases from nuclear energy systems (ENV-PE)" – Report of the collaborative project environment/potential of the international project on innovative nuclear reactors and fuel cycles (INPRO). IAEA-TECDOC, Vienna (in preparation)
3. International Atomic Energy Agency (1996) Procedures for conducting probabilistic safety assessments of nuclear power plants (Level 3): off-site consequences and estimation of risks to the public, IAEA safety series no. 50-P-12. International Atomic Energy Agency, Vienna
4. Jow HN, Sprung JL, Rolstin JA, Ritchie LT, Chainin DI (1990) MELCOR accident consequence code system (MACCS). NUREG/CR-4691. SAND86-1562, vol 2, Sandia National Laboratory, Albuquerque, NM 87185
5. Kimura M, Ishikawa J, Homma T (2009) Consideration of off-site emergency planning and response using probabilistic accident consequence assessment models. In: Proceedings of the National Radiological Emergency Preparedness Conference (NREP) conference, Landing, NJ, 22 April 2009. Available in http://www.nationalrep.org/presenters_main/Inventory2009.aspx
6. Taylor NP (2007) Wolfgang Raskob updated accident consequence analyses for ITER at Cadarache. Fusion Sci Technol 52:366
7. ICRP (2007) The 2007 recommendations of the International Commission on Radiological Protection ICRP publication 103. Annals of the ICRP 37(2–4)
8. Nuclear Regulatory Commission (U.S.) (NRC). NUREG/CR-7110 (2012) State-of-the-art reactor consequence analyses project volume 2: Surry integrated analysis. Prepared by: Sandia National Laboratories Albuquerque, New Mexico, January 2012
9. EPA (2002) Federal Guidance Report No.13 CD supplement, EP 402-C-99-001, Rev.1, April 2002. Prepared by Oak Ridge National Laboratory, Oak Ridge, TN for office of air and radiation, U. S. Environmental Protection Agency, Washington, DC
10. International Atomic Energy Agency (2008) Guidance for the application of an assessment methodology for innovative nuclear energy systems. INPRO manual, IAEA-TECDOC-1575, IAEA, Vienna
11. BOE-A-2009-17889 BOLET´iN OFICIAL DEL ESTADO Núm. 271 Martes 10 de noviembre de 2009 Sec. I. Pág. 94301
12. Mancini FCV, Gallego E, Ricotti ME (2014) Revising the emergency management requirements for new generation reactors. Prog Nucl Energy 71:160–171
13. RODOS home page (2014) Available at: http://www.rodos.fzk.de/rodos.html. Cited August 2014
14. Dvorzhak A, Puras C, Montero M, Mora JC (2012) Spanish experience on modeling of environmental radioactive contamination due to Fukushima Daiichi NPP accident using JRODOS. Environ Sci Technol 46:11887–11895
15. Andronopoulos S, Davakis E, Bartzis JG (2009) RODOS-DIPCOT model description and evaluation. Report RODOS(RA2)-TN(09)01. Available in http://www.rodos.fzk.de/rodos.html
16. Thykier-Nielsen S, Deme S, Mikkelsen T (1999) Description of the atmospheric dispersion module RIMPUFF RODOS(WG2)-TN(98)-02
17. Giménez M, Caputo M (2014) PSA level 3: test 1 source term description. IAEA/INPRO-ENV-PE 3rd meeting, April 2014. http://www.iaea.org/INPRO/CPs/ENV-PE/3st_consultancy
18. Barón JH, Chiossi CE, Vallerga HR (199) PSA-based regulatory approach in Argentina. In: Proceedings of the international conference on probabilistic safety assessment and management (PSAM 4), New York, 13–18 September 1998

Chapter 22
Novel Approaches to Radiation Diagnostics for Population Screening and Incident Triage

Michael R. McNeely

Abstract Capability deficits or capability gaps exist in the tools available for responding to and managing significant radiological incidents, whether they are power plant disasters, dirty bombs, or nuclear detonations. The lack of important response capabilities can ultimately lead to increased suffering, increased mortality, and extended recovery periods. This paper discusses the perceived gaps, requirements for products that could fill these gaps, and GattaCo's efforts in product development to meet these product requirements. These include products suitable for patient triage in the immediate aftermath of an incident and inexpensive, disposable radiobioassay test kits for population screening and environmental monitoring for long term response management.

22.1 Introduction

A significant radiological event may be accompanied by a breakdown in local infrastructure and logistics capabilities. If infrastructure is not compromised, regional capacity for diagnosing and treating radiological involvement in injuries can be easily overwhelmed with patient samples likely running in the thousands or more. In addition, long term monitoring of populations affected by radiological releases, or ongoing monitoring of environmental contamination, can become extremely expensive and burdensome. This is particularly true if comprehensive monitoring is performed which includes the analysis of biological and environmental samples for the presence of pure or near pure Alpha or Beta emitters. These radioisotopes cannot be easily detected or identified using Gamma spectrometry alone.

M.R. McNeely (✉)
GattaCo LLC, Murrieta, CA, USA
e-mail: michael.mcneely@gattaco.com

S. Apikyan, D. Diamond (eds.), *Nuclear Terrorism and National Preparedness*, NATO Science for Peace and Security Series B: Physics and Biophysics, DOI 10.1007/978-94-017-9891-4_22

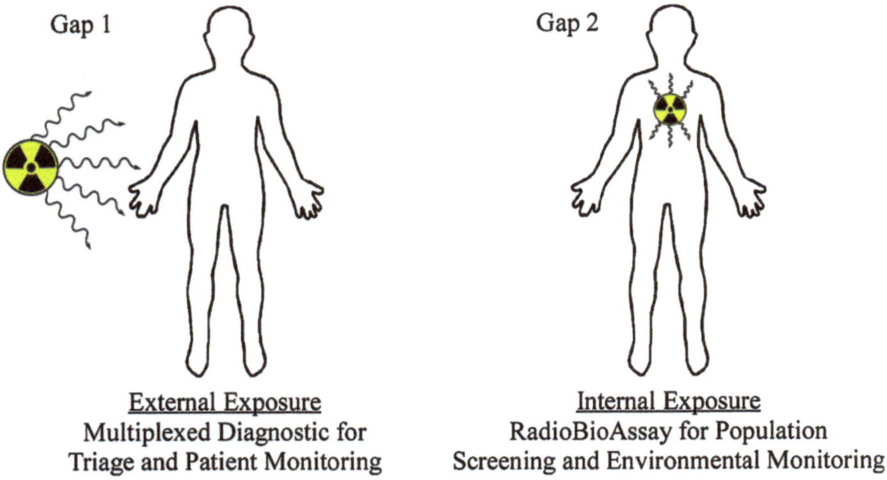

Fig. 22.1 Illustration of the capability gaps

There is a gap in testing capability in these two critical areas, as illustrated in Fig. 22.1.

22.2 Capability Gaps

22.2.1 PON MxDxEx

The first capability gap is the need of an acute to near term (<1 week) diagnostic of patients potentially exposed to a significant external radiation source. It would fill the need of triaging and monitoring when near-term therapeutic decision making is required, before patients or patient samples can be transported to advanced facilities and before obvious signs of acute radiation syndrome become apparent or when the symptoms that are expressed are complicated by physical injuries or other medical conditions.

The type of product that could fill this gap would need to be capable of detecting multiple biomarkers simultaneously including biomarkers that have been identified as being indicators of radiation exposure, and biomarkers indicative of other conditions. The purpose would be to rule-out or rule-in different conditions or syndromes to form the basis of therapeutic intervention, not to estimate some esoteric notion of exposure dose.

Ideally this product could be used in resource limited settings with no or minimal training for performing the test and interpreting the results. For patient monitoring the diagnostic would need to be performed multiple times over a 24 h to 1 week period when data logging and data transmission would be helpful.

This product is referred to as a multiplexed (Mx) diagnostic (Dx) for patient triaging and monitoring (sometimes referred to as Point of Care or Point of Need, or PON), specifically for detecting exposure to significant external (Ex) radiation sources. A useful acronym for this product category is PON MxDxEx.

22.2.2 *Radiobioassay*

The second capability gap is in ongoing population and environmental screening. Different than the PON MxDxEx device, a radiobioassay looks for radioactivity in biological or environmental samples, not biomarkers caused by radioactivity. When an area or region has been contaminated with radioactive debris it is highly likely that some of that debris makes its way into the local food chain and becomes ingested by the population. While it is relatively easy to detect low-level Gamma radiation with a simple G-M meter or Gamma spectrometer, many significant radionuclides can be missed using these devices. Specifically, pure or near pure Alpha or Beta emitters are not easily detectable with these systems because they have no significant Gamma emission. Only very sophisticated radiation detection instruments are capable of detecting Bremsstrahlung which may accompany pure Alpha or Beta emission, and it is unlikely such systems would be available for widespread use. A G-M meter or Gamma spectrometer may be capable of detecting some pure Beta emitters, but only if used very carefully and only if they are broadly available for very long testing periods of multiple hours per sample. Important pure or near pure Alpha and Beta emitters include H-3, Sr-89, Sr-90, Y-90, Cs-135, U-238, Pu-239, Pu-241, and Po-210.

Due to the complicated sample processing requirements associated with the detection and identification of pure or near pure emitters, the time period associated with data generation of a single sample can be 2–3 days or longer at a cost in the hundreds of dollars per sample. When thousands of samples are presented, data can take months to produce at a cost of millions of dollars, delaying important therapeutic countermeasures or environmental remediation efforts.

The type of product that could fill this gap is one that could screen samples for potential contamination to reduce the burden placed on labs for performing more complex analysis. Such a product could reduce complex or comprehensive sample analysis needs by 50–90 % depending on the nature of the radiological incident. The device should be capable of distinguishing between natural and man-made radioactivity, and be capable of distinguishing naturally occurring radionuclides at normal to elevated concentration levels.

Similar to the PON MxDxEx, ideally this product could be used in resource limited settings with no or minimal training for performing the test and interpreting the results. A test that detects or analyzes radioactive contamination in biological or environmental samples is sometimes referred to as a radiobioassay. For the purpose of the radiobioassay under discussion biological samples include urine, breast milk, blood, salt water and food products of animal or plant origin. Environmental samples include fresh water, air and soils with low salt content.

22.3 Filling the Capability Gaps

22.3.1 *MQ-LF™ Solution for the PON MxDxEx Capability Gap*

Numerous biomarkers have been shown to be associated with exposure to radiation [1–4]. These biomarkers are hematological, genetic (cellular), salivary and gastrointestinal based and can be detected in blood, plasma, urine, tissue and saliva samples. Unfortunately all or most of these biomarkers are also associated with non-radiation based illnesses or conditions, such as stress, anemia, hepatitis, leukemia, appendicitis, food poisoning, cardiac distress, and blunt force trauma. Many of these conditions could be present in patients also suspected of exposure to radiation.

As has been recognized the solution to distinguishing between radiation and non-radiation based biomarker elevation is to multiplex, or measure multiple biomarkers simultaneously. Ideally the measured biomarkers would span multiple syndrome pathways and include biomarkers more conclusively associated with non-radiation based conditions to aid in ruling-out radiation injury. If the biomarkers measured indicate exposure they should also be linked to acceptable therapeutic decision making.

In a resource limited setting, such as in a pre-hospital or emergency department environment where patient triaging may take place, some diagnostic testing methods are simply not practical, are too complicated, or require access to facilities and equipment that may not be available. These include detection of genetic biomarkers. Some sample types may also be difficult to acquire, such as a saliva or urine sample from a heavily traumatized or unconscious patient. A blood sample, from a finger prick to a venous draw may be more practical. Many relevant biomarkers indicative of radiation exposure or other conditions are available in blood or plasma.

A ubiquitous diagnostic platform well known for its ease of use is the lateral flow immuno-assay (LFIA) or lateral flow test strip. A common example is the home-use pregnancy test strip. Lateral flow strips are normally used for single target or single biomarker detection in qualitative, or yes or no, applications. They are less well suited for quantitative or exact measurements and particularly not well suited for quantitative measurements of multiple targets, or multiplexed and quantitative applications (MxQuant or just MQ). However, the reasons they are not well suited for MQ applications are well known and well studied. They are associated with the irregular and non-repeatable characteristic associated with poor sample volume control, poor reagent interaction and mixing control, and non-uniformities associated with sample and reagent flow through the membranes normally used in their construction.

GattaCo is developing new tools designed to eliminate the weaknesses of lateral flow to make it capable of the MQ applications under discussion. These new tools are also designed for ease of use and, in conjunction with a modified lateral flow design, provide accurate and prompt measurement of a number of relevant biomarkers available in blood in less than 15 min. Additional components available from collaboration partners include integrated real-time digital read-out of test results, data logging, data transmission, and expert systems providing likely diagnosis and treatment recommendations.

The new tools under development by GattaCo include a cartridge capable of separating plasma from whole blood to replace the error prone plasma separation membranes currently used on lateral flow strips. This cartridge is called the DEB-F™ (for DEployable Blood Filter). At present it is the only known method of plasma separation that does not require external equipment, power or multiple processing steps on the part of an operator. Just add a sufficient amount of whole blood, normally from a blood collection tube, to the inlet of the cartridge and collect the separated plasma from the outlet.

The plasma is collected from the DEB-F using GattaCo's second new tool under development, the SILO™. The SILO (for Sample In, Label, sample Out) is a precision disposable pipette designed to extract a precise volume of plasma from the DEB-F, mix it with a multiplexed biomarker labeling reagent stored inside the SILO during manufacturing, and dispense an exact volume of the mixture back out onto a modified and optimized lateral flow strip. The modified strip and associated tools are known as the MQ-LF platform. Together they are illustrated in Fig. 22.2.

As the multiplexed test proceeds, the process and ultimate results are monitored and read by an integrated digital strip reader developed by Z-integrated Digital Technologies, or ZDT Inc.. ZDT already has integrated digital strip readers commercially available in two over-the-counter products. An example of ZDT technology is shown in Fig. 22.3.

For data logging, data transmission, and expert analysis, diagnosis and treatment recommendations, iAssay® Inc. is developing the iAssay multi-format test reader and smartphone interface. Whether working solely with GattaCo's MQ-LF and integrated ZDT reader, or with other compatible and complementary products, the iAssay provides a significant data logging, data processing and wireless communications capability to connect a resource limited setting to the outside world, or to an on-board AI or expert software program, or App, pre-loaded on the device. An early prototype of the iAssay is shown in Fig. 22.4.

While the current efforts in the development of the MQ-LF platform will provide substantial multiplexed and quantitative diagnostic capability to resource limited settings, a new generation of lateral flow replacement technologies under

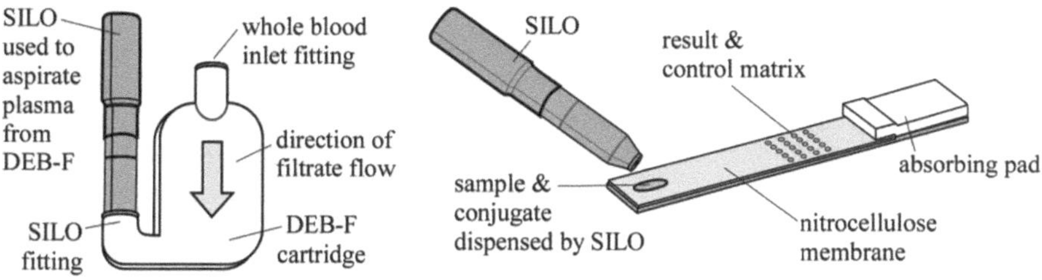

Fig. 22.2 MQ-LF platform showing SILO & DEB-F tools and modified lateral flow strip

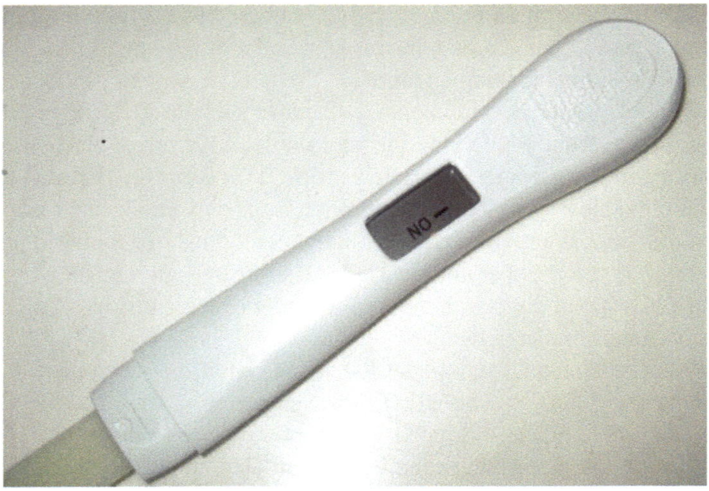

Fig. 22.3 An example of integrated reader technology

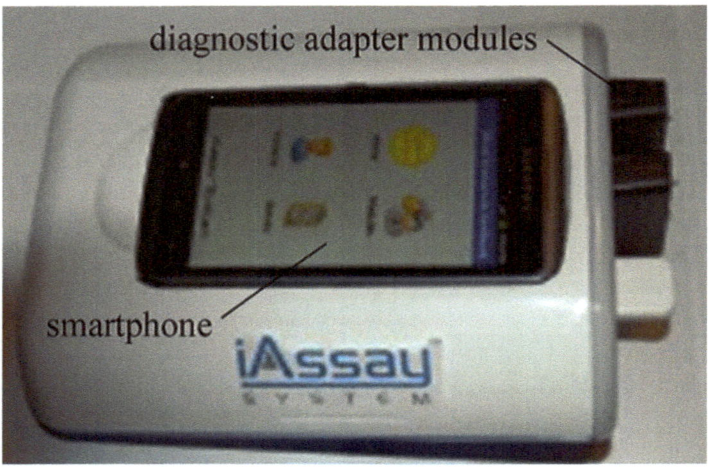

Fig. 22.4 iAssay® diagnostic cartridge and smartphone interface

development by multiple companies, has the potential of even more significant improvements. GattaCo technology is also key to the success and applicability of these new technologies to the needs of PON MxDxEx. For example, in next generation lateral flow the common but non-uniform and error prone nitrocellulose membrane is replaced with a single or multiple uniform microchannels formed in plastic. Reagents are printed in the channels using ink-jet printing technology ensuring repeatable and reliable interaction with flowing plasma, and multichannel configurations ensure multiplexing is handled by multiple individual reactions rather than in a bulk format, as is currently the case.

GattaCo's contribution to the new generation technology is the ability to separate a precise volume of plasma from whole blood and deliver it directly to the microchannel format, rather than working with the external DEB-F and SILO products. In this case the blood sample is from a finger-prick rather than a venous draw, making the task of running the diagnostic even easier. As before, an integrated digital result display and associated data logging, analysis and transmission are possible.

22.3.2 EZRADCheck™ Solution for the Radiobioassay Capability Gap

When a significant radiological release takes place tools need to be immediately available for understanding the breadth and depth of the incident, in order to best manage and mitigate its adverse effects. New inexpensive colorimetric dosimeters fill a huge capability gap that once existed. Despite their value, they are not sensitive enough, by about three orders of magnitude, to act as radiobioassays, for which they were not intended or designed. However, they serve as a very good model for the type of device needed to fill a still existing capability gap, a device that can act as a radiobioassay for screening environmental and biological samples to provide rapid information about the level of radioactive contamination present, specifically for hard to detect radionuclides, either for immediate action or to pass the suspect sample on for more detailed analysis. A qualifying device should be easy to use, inexpensive, and capable of broad spread distribution throughout an affected area.

Cursory assessment of the presence of pure Alpha or Beta emitters in the environment is not sufficient to fully understand or appreciate the dangers they represent. The most significant pure Beta emitter that is constantly overlooked or dismissed, yet is of considerable biological significance, is the long half-life radioisotope of Strontium, Sr-90. Focusing solely on the presence of easy to detect Gamma emitters, specifically Cs-137, is also not sufficient because of the vastly different biological pathways they follow. As is commonly known, Cesium mimics Potassium to a certain degree in biological organisms, including humans, and is distributed more or less uniformly throughout the organism. Cesium, as Potassium, is actively regulated by the organisms and, in humans does not tend to concentrate in one region and has a biological half-life of around 70 days. Sr-90, on the other hand, mimics Calcium and is readily taken up in the bones and teeth of contaminated individuals. Its biological half-life is roughly 28 years. Thus, if an individual is removed from the contamination source, radio-Cesium will eventually be eliminated, whereas radio-Strontium will remain in the individual for decades.

Due to the effects of bioconcentration and biomagnification, and the very long biological half-life of Sr-90, the concentration of Sr-90, as it moves up the food chain, would steadily increase. Studies have shown that the concentration of Sr-90 in the bones of fish can be more than four orders of magnitude greater than in the waters that surround them [5, 6]. As those fish are eaten by other fish, and eventually by humans, the concentration of Sr-90 could be substantially higher. Considering further a situation where radioactive contamination is periodically passed through a region, such as by periodic dumping or release of built-up nuclear reactor water into an aquatic environment, but between contamination events the surrounding waters may return to normal low levels of background radioactivity. If an organism higher up in the food chain were analyzed, the level of Cs-137 would likely be low, if enough time has elapsed to allow for the elimination of any Cs-137 that was consumed, such as 6 months or longer. Under some circumstances the fish could be allowed for human consumption and measurement of Sr-90 would not take place. But in this case the level of Sr-90, built up by bioconcentration and biomagnification and the lack of elimination of the incorporated radionuclide over the 6 month period, could in fact be thousands of times higher than any Cs-137 that were measured, and would never be allowed for human consumption if the measurement of Sr-90 ever took place.

The same scenario regarding fish and a changing aquatic environment could be said of animals that graze or migrate through a region that has a heavy contamination load, even if the region is just a hillside that may have experienced a heavy deposition due to its position and orientation along the path of a radioactive gas plume. Gamma emissions should never be used as surrogates for an estimation of pure emitters, especially in the case of Sr-90 and especially in the case when the purpose of testing is to consider whether the test subject is fit for consumption.

One significant challenge in the development and use of a radiobioassay is how to take into account the presence of naturally occurring radionuclides, specifically K-40 but also H-3, C-14 and naturally occurring radioisotopes of Uranium, Thorium and others. In its natural abundance in the environment, H-3 can contribute as much as 0.6 Bq/L to a radiobioassay measurement [7]. The presence of heavy metals, such as Uranium and Thorium, should not be deducted from a measurement because, despite their natural occurrence, they pose a significant threat to human health both because of their radioactivity but also because of the toxicity associated with heavy metals. C-14, in its natural abundance, would likely not represent a significant interferent in a radiobioassay measurement. However, due to the strong Beta signature of K-40, and its natural abundance in biological samples, which tend to have a high concentration of Potassium compared to the environmental samples, the contribution of K-40 to a measurement of radioactivity in a biological sample can be significant [8–10].

Another challenge associated with following the model of an inexpensive, disposable plastic dosimeter, especially in the case of developing and using an ultrasensitive device as is required if it is to be used as a radiobioassay, is: If it is ultra-sensitive to radioactivity, and if it is a one-time use device, how do you make it such that it measures radioactivity from a sample being tested and not respond to radioactivity coming from a significant nearby radiation source? It is impractical to require the device to always be stored and used in a heavily shielded environment. The answer is that the device should only be sensitive to Alpha and Beta radiation, and not Gamma or Neutron or X-ray emissions.

As is well known, Beta particles do not travel significant distances, generally only a few mm in water depending on their initial energy, and Alpha particles travel less than 0.1 mm in a similar environment [11]. Making the product so that it is sensitive only to Alpha and Beta and not photon emission solves the photon interference issue but also adds additional challenges, specifically in the physical design and structure to accommodate such short particle path lengths.

Also, for the device to be useful it must be easy to operate but also sensitive enough to provide relevant actionable results over a reasonable test period. As a goal we selected a target sensitivity of around 1 Bq/L over a counting period not to exceed 2 h. As has been mentioned, the device should also provide results that are easy to interpret by non-expert users, especially for samples that should not have significant interference from K-40, which, in general are environmental samples rather than biological samples. For biological samples accommodation must be made for the presence of high concentrations of Potassium, which may require the device to be used by more skilled operators such as in a clinic setting.

To accommodate the device requirements a design was created that uses ionic-triggering of self-propagating polymerization as a detection mechanism. The candidate monomer is stored in multiple individual cells comprising a matrix or grid which is located on two sides of a sample chamber within a cartridge device. As Alpha or Beta particles are emitted, they penetrate into a detection cell triggering polymerization within that cell which propagates until it fills the cell and produces a visible result, such as a color change. After an appropriate amount of time the number of cells that have changed color are counted and correlated to an approximate concentration of radioactive contamination within the sample.

For environmental samples with normally low levels of Potassium, interference from K-40 is not expected to be an issue. Due to path-length issues, detection of normal levels of low Beta particle energy radionuclides, such as H-3 and C-14 would probably not be seen, although artificially high levels may be detectable. Also, due to path length issues, detection of Alpha particles is not possible in 1st generation devices, but possibly in later generation products. For biological samples a separate measurement of Potassium concentration in the sample, such as by commonly available ionic concentration meters for liquids, may need to take place, with the amount of K-40 present in the sample calculated from the measured Potassium concentration and deducted from the measured result. Due to the nature of ionic-triggering to initiate a polymerization reaction, rather than free-radical initiation, low levels of photonic interference should not affect the results. Ionic triggering is sensitive to the interaction of Alpha or Beta particles, which are charged species, with the stored monomer.

To use, a liquid sample would be loaded into the sample chamber through an easy to access inlet port. In an alternative design the cartridge could be opened and a sample, such as an air filter, could be loaded into the sample chamber. The spacing between sheets is kept as thin as possible, such as 1 mm or less, to allow for maximum capture of Beta particle energy. The cells are visible from the outside of the device and the triggered cells are counted after an appropriate time period. The EZ**RAD**Check cartridge concept is illustrated in Fig. 22.5.

The purpose of the kit is to identify the presence of unusual Beta particle activity. If no unusual activity is detected, the conclusion can be drawn that no significant

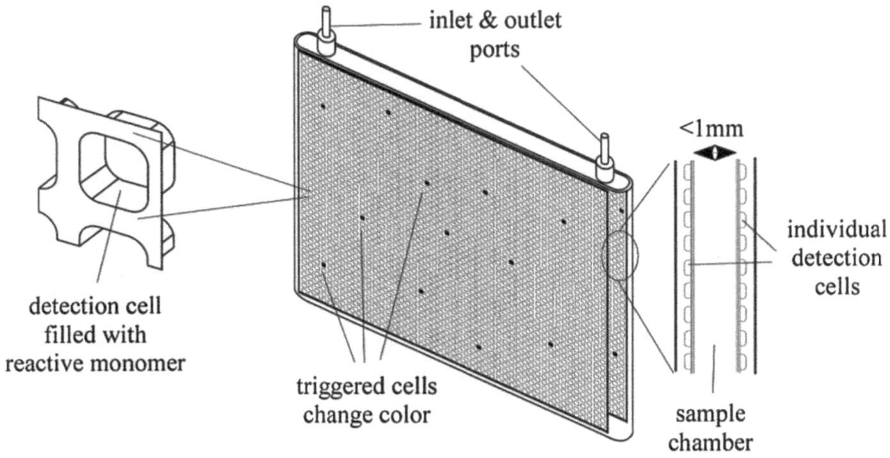

Fig. 22.5 EZ**RAD**Check kit design concept

contamination is present. If contamination is detected the sample can be passed on for more detailed analysis. The goal is to eliminate or reduce the burden placed on labs for performing the more expensive and time-consuming, yet substantially more detailed, analysis that has and will accompany a significant radiological incident.

22.4 Summary

The purpose of the EZ**RAD**Check and MQ-LF devices are to fill gaps in capabilities needed to properly manage and mitigate the impact of significant radiological incidents. The MQ-LF is for medical management in triage and short-term patient monitoring in resource limited settings. The EZ**RAD**Check is for longer term population screening and environmental awareness where current solutions and monitoring methods are limited or very expensive and time consuming.

Although both products are still under development, once completed and commercially available, they will fill a need to guide short-term therapeutic decision making and long-term environmental remediation efforts that can save lives and improve outcomes in otherwise very disagreeable circumstances.

References

1. Blakely WF (2001) Radiation exposure assessment using cytological and molecular biomarkers. Radiat Prot Dosim 97(1):17–23
2. Durante M (2003) Potential applications of biomarkers of radiation exposure in nuclear terrorism events. Physica Medica 19(3):191–212

3. Blakely WF, Miller AC, Muderhwa JM, Manglapus GL, Leidel JM, Martinez M, Landauer MR, Grace MB, Prasanna PGS (2005) Development and validation of radiation-responsive protein bioassays for biodosimetry applications. Armed Forces Radiobiology Research Institute, Bethesda
4. Guipaud O, Benderitter M (2008) Protein biomarkers for radiation exposure: towards a proteomic approach as a new investigation tool. Annali dell'Istituto superiore di sanitÃ 45(3):278–286
5. Vanderploeg HA, Parzyck DC, Wilcox WH, Kercher JR, Kaye SV (1975) Bioaccumulation factors for radionuclides in freshwater biota. Oak Ridge National Lab, Oak Ridge
6. Cummins CL (1994) Radiological bioconcentration factors for aquatic, terrestrial, and wetland ecosystems at the Savannah River Site. Westinghouse Savannah River Co., Aiken. Funding organisation: USDOE, Washington, DC
7. Mook WG, DeVries JJ (2000) Volume I: Introduction: theory methods review. In: Environmental isotopes in the hydrological cycle – principles and applications, International Hydrological Programme (IHP-V), vol 39, Technical documents in hydrology (IAEA/UNESCO). International Atomic Energy Agency (IAEA), Vienna
8. Papastefanou C, Manolopoulou M, Sihletidis L (2004) Radioactivity in human tissues. In: 11th international congress of the international radiation protection association, Full paper, Madrid, 23–28 May 2004
9. Risica S, Rogani A, Tancredi F, Grisanti A, Grisanti G, Baronciani D, Zanini R (1996) Caesium transfer to placenta, urine and human milk. In: One decade after Chernobyl: summing up the consequences of the accident, 8, 215
10. Rabitsch H, Pichl E, Kahr G, Pletz J (1996) Activity levels of Cs-137 and K-40 in the gastrointestinal tract of a cow. Radiation protection in neighbouring countries in Central Europe-1995, 194, 201
11. Diehl JF (2004) Handbook of radioactivity analysis. Edited by Michael F. L'Annunziata

Chapter 23
Radiation Injury Treatment Network®: A Model for Medical Preparedness for a Mass Casualty Radiation Incident

Cullen Case Jr. and Curt Mueller

Abstract The threat of a possible radiological disaster resulting in mass casualties continues to be on the forefront of emergency planners minds. The medical community will undoubtedly be taxed by the resulting medical surge. The United States has a well-defined emergency medical system to respond to such incidents, the National Disaster Medical System, however one area that is unique for radiological disasters is the care for casualties with Acute Radiation Syndrome. Hematologists and Oncologists purposefully expose their cancer patients to high doses of radiation as they treat their patient's cancer; resulting in symptoms identical to casualties from a radiological disaster or terrorist attack. This makes the staff from cancer centers ideal for the specialized care that will be required for thousands of casualties following a mass casualty radiological incident.

To meet this need the Radiation Injury Treatment Network® (RITN) was formed in 2006. US based RITN is a model for how a collaborative effort can fill a readiness gap with minimal effort. Through its' network of 72 medical centers, blood donor centers and umbilical cord blood banks, the RITN prepares to apply its expertise in the treatment and management of bone marrow failure, for the just such a tragic incident as the detonation of an improvised nuclear device (IND).

On March 25, 2014 President Obama said "I continue to be much more concerned when it comes to our security with the prospect of a nuclear weapon going off in Manhattan" [1]. This was in response to a question about how concerned he was of what was happening in the Ukraine. He followed this statement with "Russia's actions are a problem, but they don't pose the number one national security threat to the United States" [1]. This concern about the detonation of an improvised nuclear device on United States soil is a recurring theme that the current US President and his predecessor both have held closely and vocally, the terrorist nuclear threat is real and possible and should be planned for.

C. Case Jr. (✉) • C. Mueller
Radiation Injury Treatment Network, Minneapolis, Minnesota, USA
e-mail: ccase@nmdp.org

S. Apikyan, D. Diamond (eds.), *Nuclear Terrorism and National Preparedness*, NATO Science for Peace and Security Series B: Physics and Biophysics,
DOI 10.1007/978-94-017-9891-4_23

The United States has many assets, organizations, agencies and programs preparing for the full gamut of disasters; including a radiological or nuclear disaster. Shortly after the September 11th attacks a previously identified gap began to be filled. The need for specialized supportive care for casualties with radiation only injuries from a radiological incident including acute radiation syndrome. This prompted the National Marrow Donor Program and the American Society for Blood and Marrow Transplantation to engage the Hematology and Oncology community to devise a solution to fill this gap which resulted in the formation of the Radiation Injury Treatment Network® (RITN).

The Radiation Injury Treatment Network provides comprehensive evaluation and treatment for victims of radiation exposure or other marrow toxic injuries. Many of the casualties with radiation injury will be salvageable but require outpatient and/or inpatient care. Recognizing this, the US National Marrow Donor Program, US Navy and American Society for Blood and Marrow Transplantation collaboratively developed RITN, which comprises of medical centers with expertise in the management of bone marrow failure, stem cell donor centers and umbilical cord blood banks across the US.

The US based RITN is a model for how a collaborative effort can fill a readiness gap with existing resources. Through its' network of 72 hospitals, cancer treatment centers, blood donor centers and cord blood banks the RITN is preparing for the resulting medical care for radiation only casualties from the detonation of an improvised nuclear device.

The goals of RITN are:

- to develop treatment guidelines for managing hematologic toxicity among victims of radiation exposure,
- to educate health care professionals about pertinent aspects of radiation exposure management through training and exercises,
- to help coordinate the medical response to radiation events, and
- to provide comprehensive evaluation and treatment for victims at participating centers.

One thing that differentiates the RITN from other efforts is that RITN hospitals plan to be distant to the incident. If there is a radiological disaster with mass casualties with marrow toxic injuries on the Eastern coast of the US it is anticipated that the local RITN hospitals will be overwhelmed with absenteeism, trauma casualties and caught up in the chaos of the incident and therefore unable to provide the specialized medical care that the radiation only casualties will require. RITN hospitals that are outside the disaster area will be resource rich in comparison and able to care for the radiation only casualties that are moved through the National Disaster Medical System. Following the activation of the National Disaster Medical System initially trauma casualties will be moved away for care then the radiation only casualties who would benefit from specialized supportive care at a RITN center will be moved for care. The location of RITN hospitals are shown in the map below:

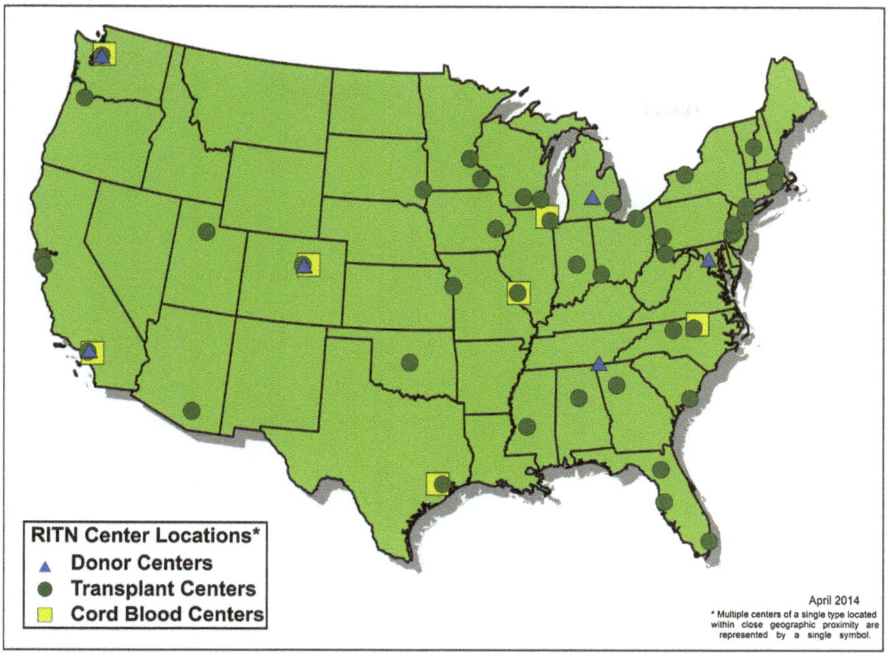

The number of potential casualties with radiation only injuries that could benefit from specialized medical care are estimated to be 63,000 [2]. These casualties would be a fraction of the total casualties (approximately 15 % [2] of the total) but the beds necessary to care for them can be available once displaced across the nation to the hospital and cancer treatment centers that compose the RITN.

Estimated Radiation Injury Only Casualties

Radiation Dose (Gy)	Care Requirement	Casualty Estimate
Mild (0.75-1.5)	Possible self recovery with limited care and some Outpatient monitoring	91,000
Moderate (1.5-5.3)	Combination of Outpatient care and Specialized Inpatient Supportive Care	51,000
Severe (5.3-8.3)	Intensive Supportive Care (most possibly including marrow transplant)	12,000
Expectant (>8.3)	Comfort Care	47,000
Combined Injury and Radiation (>1.5)	Stabilization and monitoring, pending resource availability	44,000

Estimate of 63,000 casualties for RITN

Combined Injury Kills!!

Table adapted from: Knebel et al. [2]

The number of casualties is further divided by those who would need outpatient monitoring and care vs. inpatient specialized supportive care and those who would possibly require a marrow transplant.

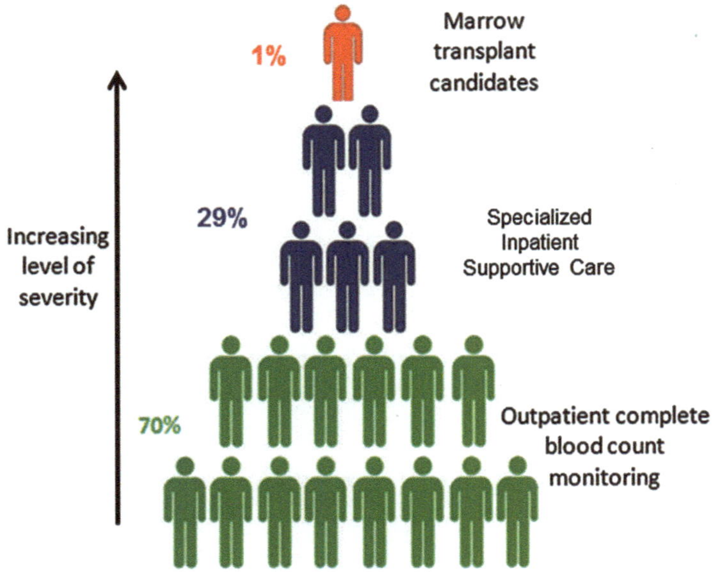

Once the breakdown of required care is applied the RITN has shown through capacity surveys that the 30 % requiring the most resource intense care will be possible. Following the Fukushima Nuclear Power Plant accident the RITN conducted a survey of its hospitals asking the number of available beds immediately under normal standards of care to how many beds would be available if applying crisis standards of care. The results were that up to 13,000 radiation only casualties could be cared for under austere conditions.

The detonation of an IND would result in a mass casualty marrow toxic incident; additionally some toxic chemicals such as Lewisite (a.k.a. Mustard) can destroy a person's marrow. Since RITNs inception in 2006 its' participants have accomplished significant undertakings to prepare for a marrow toxic injuries ranging from developing treatment guidelines, developing standard operating procedures, solidifying partnerships, conducting annual exercises to being recognized by the federal government in a recent publications as a national response asset [3–7]. The former Department of Health and Human Services Secretary Leavitt honored the Radiation Injury Treatment Network for its efforts in improving national preparedness.

One of the first steps taken by RITN was to develop standardized treatment guidelines and admitting orders for victims of a marrow toxic incident as well as to incorporate them into all RITN center SOPs. Having a standard process by which to treat a patient at a center is not enough to be prepared; testing the plan is essential to ensure all components will function cohesively. Annually RITN centers conduct tabletop exercises where all key personnel are gathered to respond to a disaster scenario (participants include a wide range from nursing, to pharmacy, admitting to social workers). They walk though step by step how they would receive the patients, verify they are decontaminated, make room for them on the already packed wards, and acquire additional resources (staff to equipment).

Being prepared is a great leap forward, but people need to know you are there to help in response to a catastrophe. Increasing awareness of RITN with-in the medical and emergency response communities is paramount to RITNs effectiveness. This was accomplished through partnerships, education and increased awareness through publications. Partnerships were established with ASBMT, AABB, Dept. of Health and Human Services-Asst. Secretary for Preparedness and Response and international organizations including EBMT and International Atomic Energy Agency-Radiation Emergency Medical Preparedness and Assistance Network. The RITN Executive Committee has diligently worked to publish articles related to marrow toxic injuries and incident response [8–10]. Over 10,000 medical professionals and their support staff have received training or education connected to RITN between 2006 and 2013 (2014 numbers are pending); ranging from a web based Basic Radiation Training course to instructor led medical grandrounds. The breakdown of the training conducted is shown in the following table:

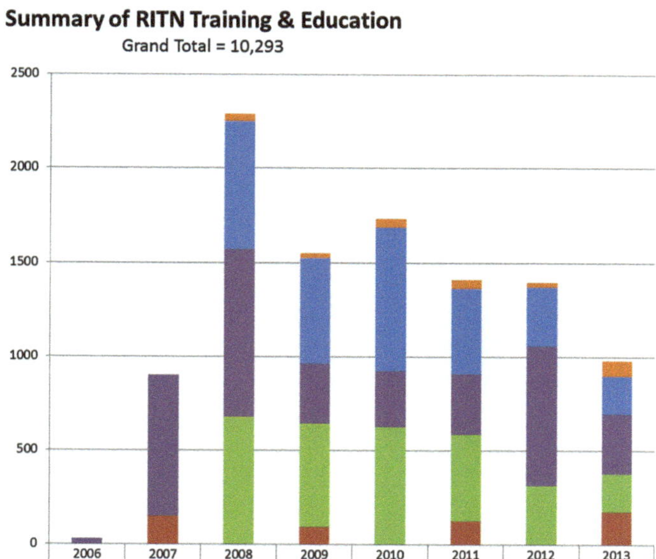

	2006	2007	2008	2009	2010	2011	2012	2013
REAC/TS			38	26	46	45	26	84
Grandrounds			677	564	763	460	314	203
Basic Radiation Tng	30	752	895	320	301	322	744	316
Overview Presentation			677	549	623	460	314	203
Conference Attendees		149		92		125		175

Another key means of preparing has been through the application of preparedness exercises. Since 2006 RITN centers have conducted 435 exercises of varying intensities. Each year RITN centers are required to conduct a RITN developed tabletop exercise where staff review a scenario and discuss in a conference room setting how they would respond then submit responses to key questions. All of these exercises as well as the results are posted on www.RITN.net/exercises. Beginning in 2013 we incorporated a web-based version of the tabletop exercise to allow hospitals from across the nation to collaborate on their responses to the scenario. Additionally, since 2012 RITN has funded the creation/development/execution of three full scale exercises where the hospital emergency operations center is activated and staff actually practice receiving casualties, conducting radiological survey, medical triage and create treatment plans. The breakdown of the exercises conducted is shown in the table below:

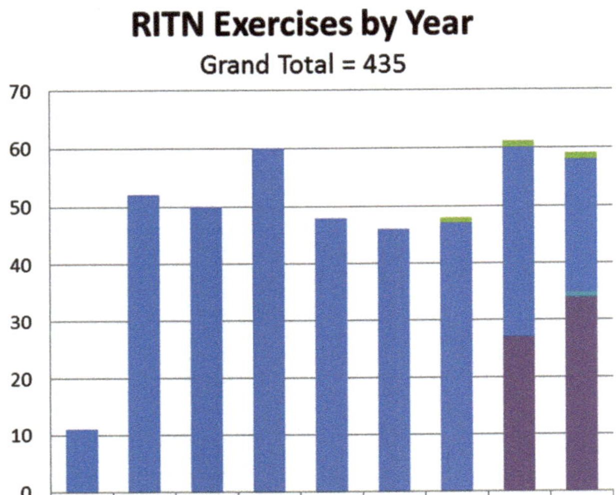

	2006	2007	2008	2009	2010	2011	2012	2013	2014
Fullscale							1	1	1
Tabletop	11	52	50	60	48	46	47	33	23
Regional Tabletop									1
Web-based Tabletop								27	34

The Radiation Injury Treatment Network has shown much promise since its inception filling a gap in US preparedness in a unique way using an existing resource that had previously been overlooked. RITN continues to expand and refine its capabilities to be better prepared for a mass casualty radiological disaster with marrow toxic injuries.

RITN Mission The Radiation Injury Treatment Network® (RITN) provides comprehensive evaluation and treatment for victims of radiation exposure or other marrow toxic injuries. RITN develops treatment guidelines, educates health care professionals, works to expand the network, and coordinates situation response. RITN is a cooperative effort of the National Marrow Donor Program® (NMDP) and The American Society for Blood and Marrow Transplantation (ASBMT).

References

1. USA Today, Obama's worry: a nuke in Manhattan. http://www.usatoday.com/story/theoval/2014/03/25/obama-netherlands-nuclear-summit-romney-manhattan-new-york-nucear-attack/6868489/. Accessed 17 Oct 2014
2. Knebel AR, Coleman CN, Cliffer KD et al (2011) Allocation of scarce resources after a nuclear detonation: setting the context. Disaster Med Public Health Prep 5(Suppl 1):S20–S31
3. Planning Guidance for Response to a Nuclear Detonation, 2nd Edition Planning Guidance for Response to a Nuclear Detonation, Second edition, 6/2010 (National Security Staff, Interagency Policy Coordination Subcommittee for Preparedness & Response to Radiological and Nuclear Threats)
4. State and Local Planners Playbook for Medical Response to a Nuclear Detonation (Department of Health and Human Services/Assistant Secretary for Preparedness and Response, March 2011). http://www.phe.gov/Preparedness/planning/playbooks/stateandlocal/nuclear/Pages/default.aspx. Accessed 12 Feb 2014
5. Responding to an RDD/RED Emergency: the HHS Playbook (HHS/ASPR, April 2010). http://www.phe.gov/Preparedness/planning/playbooks/rdd/Pages/default.aspx. Accessed 12 Feb 2014
6. Coleman CN, Case C Jr et al (2012) Medical planning and response for a nuclear detonation: a practical guide. Biosecur Bioterror 10(4):346–371
7. Coleman CN, Hrdina C, Bader JL et al (2009) Medical response to a radiologic/nuclear event: integrated plan from the Office of the Assistant Secretary for Preparedness and Response, Department of Health and Human Services. Ann Emerg Med 53(2):223–225
8. Weinstock DM, Case C Jr, Confer DL (2008) Response: radiologic and nuclear events. Blood 111(12):5758–5759
9. Weinstock DM, Case C Jr, Bader JL et al (2008) Radiological and nuclear events: contingency planning for hematologist/oncologists. Blood 111(12):5440–5445
10. Fliedner TM, Chao NJ, Bader JL, Boettger A, Case C Jr, Chute J, Confer DL, Ganser A, Gorin NC, Gourmelon P, Graessle DH, Krawisz R, Meineke V, Niederwieser D, Port M, Powles R, Sirohi B, Weinstock DM, Wiley A, Coleman CN (2009) Stem cells, multiorgan failure in radiation emergency medical preparedness: a U.S./European consultation workshop. Stem Cells 27(5):1205–1211. doi:10.1002/stem.16

Chapter 24
Radiation Dose Assessment by Using Lymphocyte Counts

Shaowen Hu, William Blakely, and Francis Cucinotta

Abstract Peripheral blood cell counts are important biomarkers of radiation exposure. With the successful application of a simplified compartmental modeling approach to simulate the perturbation of hematopoiesis system in humans after radiation exposure, we recently developed a HemoDose software program to estimate absorbed dose based on multi-type blood cell counts. Testing with patient data in some historical accidents indicates that either single or serial granulocyte, lymphocyte, leukocyte, or platelet counts after exposure can be robust indicators of the absorbed doses. In this work, the first week lymphocyte counts of five patients in the 2011 Bulgaria radiation accident are used to do serial points and single point calculations with HemoDose. Overall, the estimated doses are in good agreement with those evaluated with cytogenetic analysis in two independent laboratories. The program also confirms that calculation with individual reference counts can significantly increase the accuracy of this simple dose estimation algorithm. These results indicate that HemoDose can be employed as an easy-to-use and deployable biodosimetry tool for predicting the clinical severity, treatment, and survivability of exposed individuals and triaging those with minimum or no exposure, especially in a large-scale nuclear/radiological disaster scenario involving mass casualties.

24.1 Introduction

The United States must be prepared for radiological accidents and terrorist activities that may involve large number of irradiated personnel [1]. Researchers have put much effort to develop various physical and biological dosimeters to rapidly

S. Hu (✉)
Wyle Science, Technology and Engineering, Houston, TX, USA
e-mail: shaowen.hu-1@nasa.gov

W. Blakely
Radiation Biodosimetry Research Program, Uniformed Services University of the Health Sciences, Armed Forces Radiobiology Research Institute, Bethesda, MD, USA

F. Cucinotta
School of Allied Health Sciences, University of Nevada, Las Vegas, NV, USA

S. Apikyan, D. Diamond (eds.), *Nuclear Terrorism and National Preparedness*, NATO Science for Peace and Security Series B: Physics and Biophysics,
DOI 10.1007/978-94-017-9891-4_24

estimate radiation dose in mass-casualty and population-monitoring scenarios, which are essential for effective medical management and treatment of the exposed individuals [2]. Direct measurement of radiation exposure with physical devices can promptly detect radionuclide contamination in living subjects and measure the intensity of external radiation in the environment that can be used to estimate the absorbed dose and dose distribution in a particular individual. However, unfortunately radiation dosimeters are frequently not present in radiation accidents. To guide medical personnel in their clinical decisions, biological markers are usually applied to examine the radiation induced changes at different biological levels, i.e., total organism, organ systems, cell systems, cellular and subcellular levels. These indicators are observable signs and symptoms as a function of time after radiation exposure. The manifestation of these syndromes reflect the response of physiological processes at various levels coping with the effects of radiation impairment [3]. Among these, the peripheral blood cell counts are widely recognized as robust and high-throughput indicators to assess the extent of radiation-induced injury [4]. This is due to the fact that hematopoietic system is a vulnerable part of the human body to radiation damage [5, 6]. Particularly, the lymphocyte and granulocyte cells are the most radiosensitive of the blood elements, and monitoring their changes after exposure is regarded as the most practical and best laboratory test to estimate radiation dose [7].

The dose estimation algorithm through lymphocyte counts in this work is based on a simplified compartmental lymphopoietic model, which was used to simulate and interpret the experimental data of acute and chronic radiation on rodents [8, 9]. We developed methods to extrapolate the model for humans and successfully simulated clinical data of accidental human victims with wide ranges of absorbed doses and exposure scenarios [10]. With the human lymphopoiesis model, the simulated lymphocyte counts and the depletion rate constants are consistent with the widely recognized Guskova's and Goans' formulas, as can be seen from the data presented in Fig. 24.1 [10]. It should be pointed out that the Goans' method is valid only for data points within the first 48 h after exposure, beyond which the predictions significantly deviate from those of the Guskova's method, especially for cases with high dose exposure. On the other hand, the Guskova's method also requires use of lymphocyte counts within the first 9 days after exposure. Our investigation indicates that using the lymphopoiesis model does not have these time limitations, though the dose estimation with early time points after exposure is more reliable [11].

In addition to the lymphopoiesis models, recently we found the granulopoiesis, leucopoiesis, and thrombopoiesis models adapted for human consequence also can be used to estimate the absorbed dose based on the circulating granulocyte, leukocyte, and platelet counts, respectively. These consensus results stimulated us to develop a software "HemoDose" program, which uses the clinical hematological data after radiation exposure as inputs and gives out dose estimation by doing rapid calculation with the underlying hematopoiesis models [11]. We found that a wide range of clinical data points, not only in the early time window (1 or 2 days), but also in the late phase (up to 4 weeks) after exposure, can be used for dose calculation,

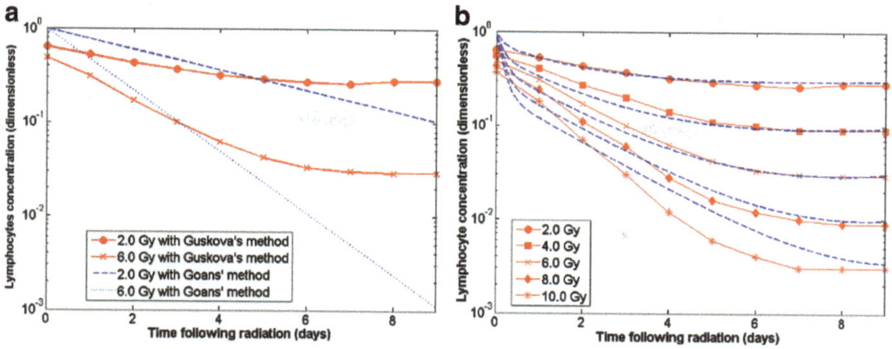

Fig. 24.1 Comparison of human lymphopoiesis model with the Guskova's and Goans' methods [10]. (**a**) The first 9 days' lymphocyte counts for humans exposed to 2.0 and 6.0 Gy, calculated by the two empirical methods. (**b**) Lymphopoiesis model outputs (*dash lines*) and the results (*lines with symbols*) of Guskova's method for the first 9 days' lymphocyte counts after acute exposure

when the four types of cell counts are combined for analysis. As the lymphopoietic model performs the best for early phase dose estimation [11], in this work we focus on the dose estimation by using serial and single lymphocyte counts of the patients in the 2011 Bulgaria accident, with comparison of the "golden standard" by cytogenetic analysis reported before [12].

24.2 Materials and Methods

24.2.1 Cases and Previous Reports

The hematological data used in this study come from the patients in the radiation accident occurred on 14 June 2011 in Stamboliyski, Bulgaria. Five people were exposed 5–10 min to ^{60}Co gamma-ray source in an industrial irradiation facility by operational mistakes. Forty hours after exposure, their prodromal signs and the dynamics of hematological parameters were examined in the National Centre of Radiobiology and Radiation Protection (NCRRP), Bulgaria [13]. Based on the estimation of exposure severity, 8 days after the accident, they were brought to the Hematology clinic of Percy Hospital in Paris, France, where treatment with growth factors was applied to help in hematopoiesis recovery. Dose estimation were conducted in two independent laboratories, one in NCRRP, another in the Intitut de Radioprotection et de Sureté Nucléaire (IRSN), Paris, with cytogenetic analysis on blood samples from each patient [12]. The age, gender, health status, as well as the estimated doses of the victims provided by two laboratories, are summarized in Table 24.1.

Table 24.1 Health information and dose estimation of the five 2011 Bulgaria accident patients

Case	Age (year)	Gender	Health status	Dose (Gy) via NCRRP	Dose (Gy) via IRSN
P1	44	F	Healthy	5.60 [4.94,6.25]	5.04 [4.40, 5.93]
P2	58	M	Healthy	3.37 [2.85, 3.89]	3.44 [2.99, 3.97]
P3	58	M	Healthy	2.47 [2.04, 2.90]	2.52 [2.17, 2.92]
P4	76	M	Diabetes and hypotension	1.96 [1.56, 2.36]	1.86 [1.59, 2.15]
P5	78	F	Healthy	1.25 [0.81, 1.70]	1.23 [0.98, 1.50]

24.2.2 Dose Estimation Algorithm Through Lymphocyte Counts in HemoDose

The lymphopoiesis model of HemoDose is used to retrospectively estimate the absorbed doses of the five patients, with the lymphocyte counts of the first week reported in [13]. These include 6 data points for Patients 1–4, and 3 data points for Patient 5. These data can be input into the Windows version of HemoDose by two ways (Fig. 24.3). The first option allows a prepared data file from a local computer to be uploaded, which contains the post-exposure times (days) and the corresponding blood cell counts (G/L) as a set of 2-column data. The times and the corresponding blood cell counts can be displayed by clicking the "View Data" button, for the user to check the correctness of input. Clicking the "Simulate" button will launch the background codes of the corresponding hematopoiesis models, which generate a best-fitting curve based upon the input data, plot the curve as well as the input data, and give out an estimated dose with uncertainty range (i.e., 95 % CI). The second option allows the user to input the clinical data manually, and, similarly, to simulate the best-fitting dynamics of specific blood cell counts as compared with the input, and to give out an estimated dose with CI (Fig. 24.2).

24.3 Results

First, all available data points for each patient are used to do serial points calculation. The results of two sets of calculation are plotted in Fig. 24.3, with comparison of the evaluation with cytogenetic analysis methods in two laboratories (NCRRP and IRSN). In HemoDose1 calculation, the default baseline (2.0×10^9 cell/L) is applied for all five patients, while in HemoDose2 calculation, the individual reference counts of lymphocyte from 20 to 26 years' health monitoring of each patient [13] are used. For Patient 3, the estimated doses by these two calculations are nearly the same, as this patient's personal reference count 1.949×10^9 cell/L [13] is very close to the default baseline of the model. The discrepancy from the averaged reported dose for this patient is about 0.8 Gy. For Patients 1, 2, and 4, the calculated doses using personal reference counts are about 0.5 Gy higher than those using default

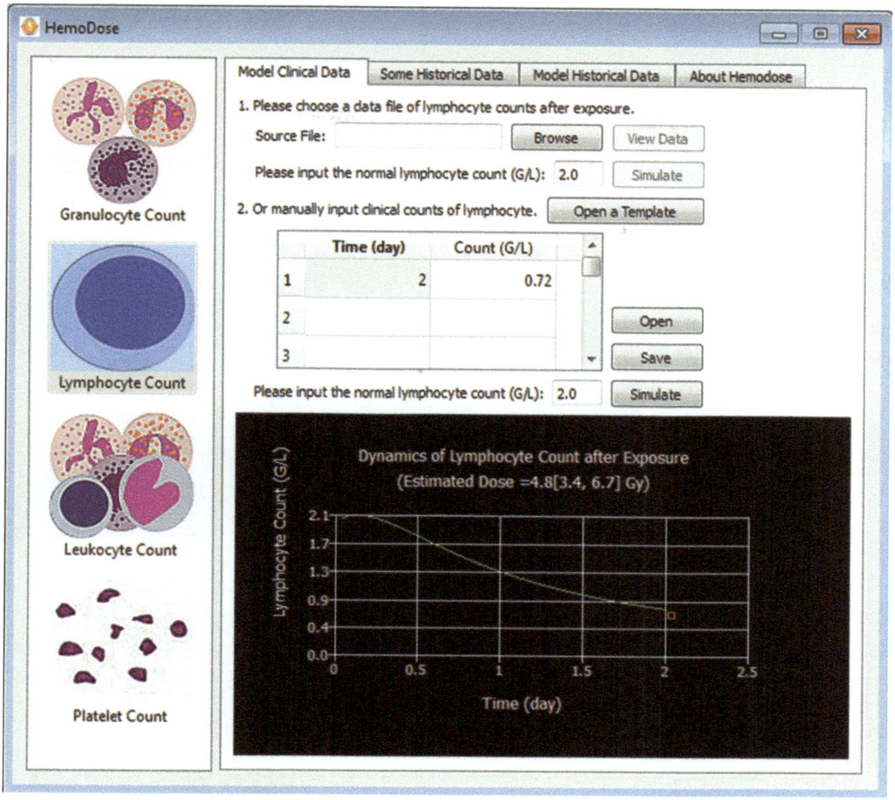

Fig. 24.2 User interface of Windows version of HemoDose to model clinical data

counts. The differences from the averaged doses evaluated by cytogenetic analysis methods are 0.2, 0.1, and 0.4 Gy, respectively (Fig. 24.1). HemoDose1 calculation for Patient 5 is about 1.7 Gy higher than the doses measured by cytogenetic methods. When using the personal reference counts, the discrepancy reduces to 0.7 Gy (Fig. 24.3). This indicates that calculation with individual reference counts can significantly increase the accuracy of dose estimation, and the estimated doses calculated by serial lymphocyte count time-points in the first week are in good agreement with those evaluated with cytogenetic analysis in two independent laboratories.

Next, every single point of lymphocyte count of the five patients is used to do single point calculation with HemoDose. The manual input option of HemoDose (Fig. 24.2) is very easy to use for this purpose. Figure 24.4 depicts the results for Patient 1, which were obtained with both the default reference count and the personal reference counts. The six doses calculated with the default baseline are reasonably close to the estimated doses by cytogenetic methods, with the exception of the Day 5 calculation which is 1.3 Gy lower than the averaged dose reported by two laboratories. When personal reference count was used, the discrepancies were

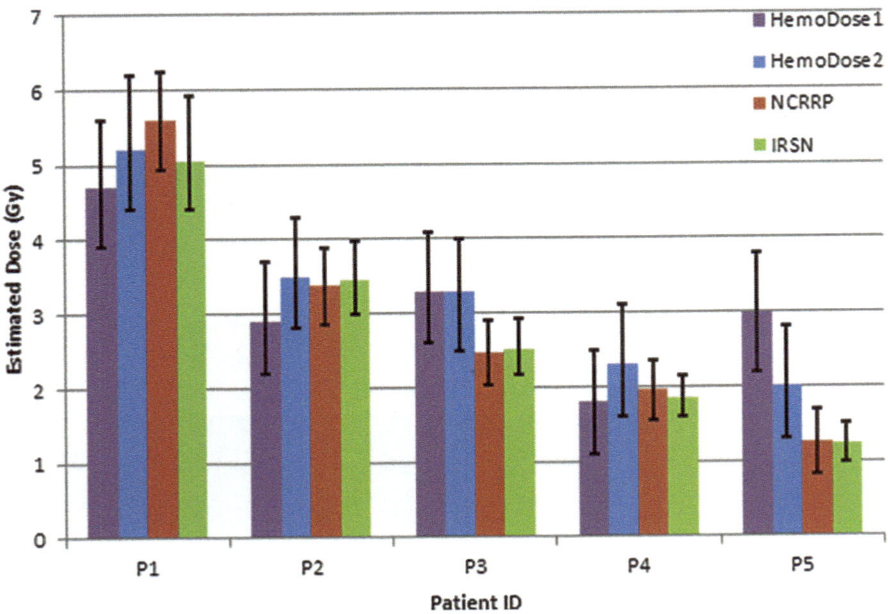

Fig. 24.3 Comparison of the estimated doses of the five patients by HemoDose and cytogenetic methods. Doses denoted as HemoDose1 are estimated with the default baseline lymphocyte count (i.e., $2.0 \times 10^9\ L^{-1}$), while those as HemoDose2 are calculated with the specific baseline count of each patient reported in [13]. *Error bars* are plotted with 95 % confidence intervals of the methods

significantly reduced for all points, with the maximum difference of 0.7 Gy for Day 6 calculation (Fig. 24.5). Overall, the estimated doses by single point calculation for Patient 1 are in good agreement with those evaluated with cytogenetic analysis in two independent laboratories.

Figure 24.5 shows the results of single point calculation for Patients 2–5. For Patient 2, all the estimated doses by two sets of calculation are reasonably close to the reported doses, with maximum discrepancy of 1.0 Gy for Day 4's calculation with the default baseline. Using the personal reference count improves the accuracy of dose estimation, as compared to the reported results. However, for Patient 3, the first three doses estimated by single point calculation are significantly higher than the reported doses (differences are about 0.9, 1.6, 0.9 Gy, respectively). This is probably related to the inherent variations in individual radiation sensitivity. Though Patients 2 and 3 are of the same age, gender, and both are free of chronic illness; it was reported that Patient 3 had diarrhea 12–14 h after the exposure [13]. Among the five victims, this is the only occurrence of such a symptom, which is generally regarded as a sign of gastrointestinal response to severe radiation [4]. Nevertheless, the other three estimated doses of this patient by single point calculation are reasonably close to the reported results, with discrepancy of 0.7, 0.2, 0.6 Gy, respectively.

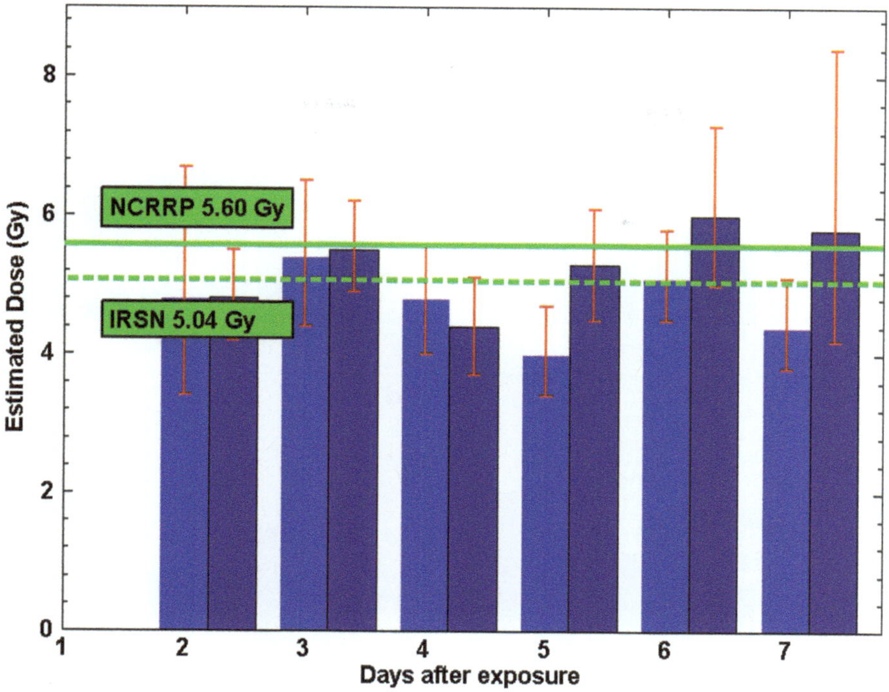

Fig. 24.4 Single point calculation of the absorbed dose of Patient 1 with HemoDose from lymphocyte counts. Doses with light *blue bars* are estimated with the default baseline lymphocyte count (i.e., $2.0 \times 10^9\ L^{-1}$), while those with *dark blue bars* are calculated with the specific baseline count of P1 reported in [13]. *Error bars* are plotted with 95 % confidence intervals of the methods. Doses estimated with cytogenetic methods at NCRRP (*solid line*) and IRSN (*dash line*) are plotted for comparison

The responses of senior patients P4 and P5 (76 and 78 years old, respectively, when exposed) of this accident can be used to verify the concern of age factor in the models of HemoDose [11]. The current version of HemoDose assumes the baseline counts are those of healthy adults (age group 18–57 years) [14]. Recent research implicated that these baselines are variant for groups of different ages and genders [15]. Particularly, the radiosensitivity parameters of the various cellular compartments for children and seniors should be different from the adults. However, in this work, the first four estimated doses of Patient 4 by single point calculation are still reasonably close to the reported doses, with either default baseline or personal baseline (Fig. 24.6). The dose estimated by Day 2's lymphocyte count for Patient 5 is also approaching the reported doses, if the personal reference count is used (Fig. 24.6). The major discrepancy is observed with Days 6 and 7's calculation for Patient 4, and Days 3 and 7's calculation for Patient 5, all of which are significantly higher than the reported doses (Fig. 24.6). Examining these counts with the simulated curves for serial time-point calculations indicates that, at these time-points, the lymphocyte counts of the seniors are lower than the predicted counts

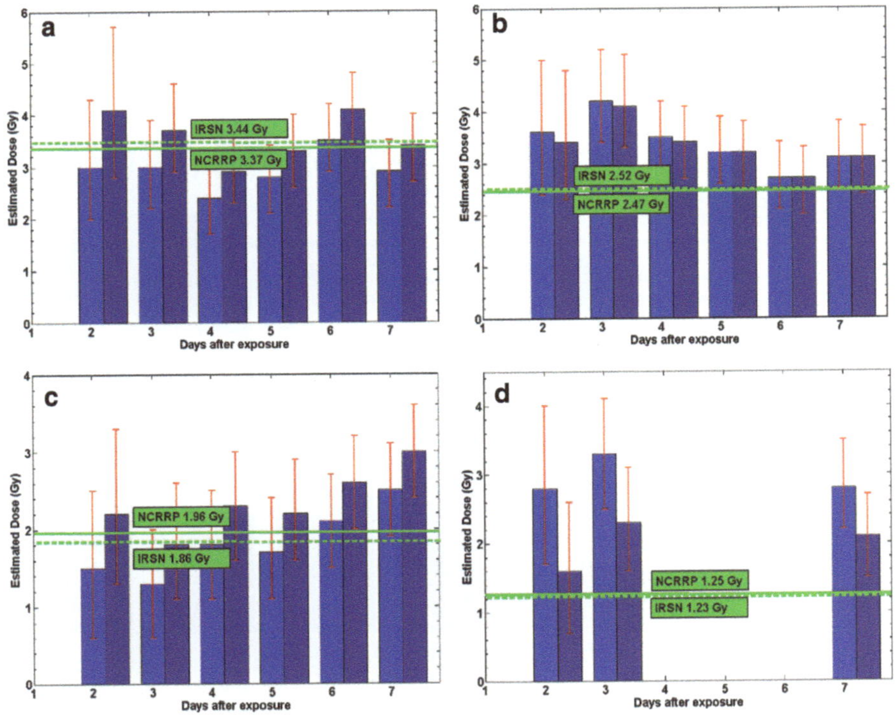

Fig. 24.5 Single point calculation of the absorbed doses of other patients with HemoDose from lymphocyte counts. Plot (**a**) is for Patient 2, (**b**) for Patient 3, (**c**) for Patient 4, and (**d**) for Patient 5. For keys see Fig. 24.5

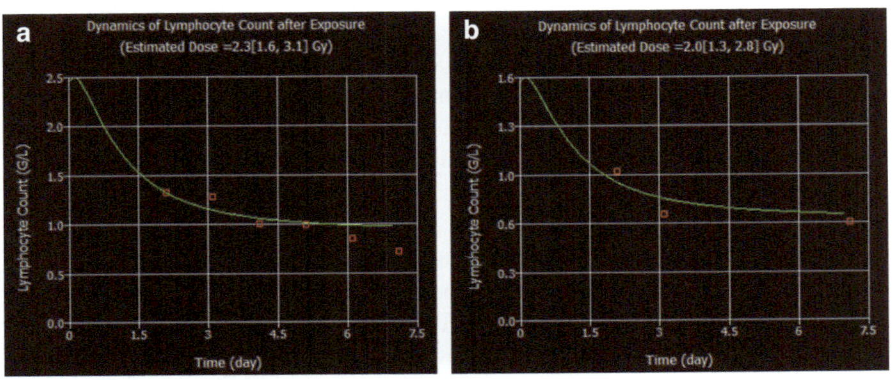

Fig. 24.6 Serial points calculation of Patients 4 (**a**) and 5 (**b**) with Windows version of HemoDose

from the model (Fig. 24.6). This higher depletion rate than the model's expectation may implicate that the lymphocytes in seniors are more sensitive to radiation than in younger aged patients. If this trend can be verified in senior victims in other accidents, it will be highly suggestive that an age factor needs to be considered in the HemoDose approach, at least with the lymphopoietic model.

24.4 Discussion

Radiation response of human body is a complicate process that involves many organs, tissues, cells, subcellular components, and biomolecules at different levels of organization. In history, many biomarkers at different biological levels have been utilized to characterize the severity of radiation injury. However, due to the inherent limitations of each method and the vast challenges in most radiation accident scenarios, it has been established that none of these methods can be used as a stand-alone tool to perform rapid and accurate dose assessment for the possible hundreds of thousands of people in such events [16]. Compared to other sophisticated methods such as the cytogenetic analysis, the HemoDose approach is inexpensive, fast, and easy-to-use, and is deployable in field. It is expected that this method can play an essential role in the initial triage during such events, and other sophisticate techniques can be used as complimentary tools to further determine the accurate doses for the victims that need medical intervention or no-exposure assurance.

Though other hematological parameters during the first week after exposure are also reported, calculation with granulopoietic, leukopoietic, and thrombopoietic models of HemoDose could not reproduce the reported doses accurately. For example, the decline in granulocyte count to reach a nadir at 3–4 weeks after irradiation if often preceded by an initial increase of granulocytes believed to be due to demargination of granulocytes and release of mature and early precursors within the bone marrow [7]. A subsequent transient rise in granulocyte may occur after exposures to moderate radiation doses between 5 and 10 days following radiation exposure. This is due to the facts that circulating granulocytes, leukocytes, and platelets are radio-resistant, and some injured cells at the stem cell level in bone marrow are still capable of performing a number of multiplicative divisions of differentiated progeny of hematopoietic progenitor cells after exposure [6]. This is assumed to be the mechanism for the appearance of an "abortive rise" of granulocytes and leukocytes, as well as a delay of decline of platelets. For these three types of cells, a serial of lowest counts can be observed usually between 20 and 30 days after exposure. It is during this period that the granulocyte, leukocyte, and platelet counts in peripheral blood are more robustly correlated with the absorbed doses [11]. Unfortunately, hematological measurement after the first week were not recorded after the patients were transferred to Percy Hospital in Paris, France, from Day 8 to Day 44 [17]. Dose estimation by either serial points or single point calculations of HemoDose with early phase counts of these cells are quite erroneous, as observed from calculations with other accidental data [11].

While the age factor may play a role in the overestimation of the doses of the two senior patients, with the lymphopoietic model of HemoDose, the gender factor seems not to be a significant contributing parameter in the dose estimation algorithm. Though all models in HemoDose are calibrated with exceptional male patients data from past radiation accidents [11], the results calculated for female patients P1, either from serial points calculation or from single points calculation, are quite consistent for male patients such as P2. To be a useful biodosimetry tool in a radiology/nuclear disaster scenario, the application range of population of these models needs to be expanded to include all segments of civil population, especial for at-risk population such as children, pregnant women, senior citizens, and patients with illness or infection, etc. Further works are necessary to consider larger accidental patients databases such as SEARCH [18] and the U.S. Radiation Registry REAC/TS [19].

Acknowledgments The opinions or assertions contained herein are the private views of the authors and are not necessarily those of the Wyle Science, Technology and Engineering, the Armed Forces Radiobiology Research Institute, the Uniformed Services University of the Health Sciences, the University of Nevada, the U.S. Department of Defense, or National Aeronautics and Space Administration (NASA). This work was supported by NASA Space Radiation Risk Assessment Project and the University of Nevada, Las Vegas. We are thankful for many years of collaboration with Dr. Olga Smirnova.

References

1. Reeves GI, Jarret DG, King GL, Blakely WF, Seed TM (1996) Proceedings: TRIAGE of irradiated personnel. AFRRI, Bethesda
2. Swartz HM, Flood AB, Gougelet RM, Rea ME, Nicolalde RJ, Williams BB (2010) A critical assessment of biodosimetry methods for large-scale incidents. Health Phys 98(2):95–108
3. Hu S, Kim MH, McClellan GE, Cucinotta FA (2009) Modeling the acute health effects of astronauts from exposure to large solar particle events. Health Phys 96(4):465–476
4. Waselenko JK, MacVittie TJ, Blakely WF, Pesik N, Wiley AL, Dickerson WE, Tsu H, Confer DL, Coleman CN, Seed T, Lowry P, Armitage JO, Dainiak N (2004) Medical management of the acute radiation syndrome: recommendations of the Strategic National Stockpile Radiation Working Group. Ann Intern Med 140(12):1037–1051
5. Goans RE, Holloway EC, Berger ME, Ricks RC (2001) Early dose assessment in criticality accidents. Health Phys 81(4):446–449
6. Fliedner TM, Graessle D, Meineke V, Dorr H (2007) Pathophysiological principles underlying the blood cell concentration responses used to assess the severity of effect after accidental whole-body radiation exposure: an essential basis for an evidence-based clinical triage. Exp Hematol 35(4 Suppl 1):8–16
7. Dainiak N (2002) Hematologic consequences of exposure to ionizing radiation. Exp Hematol 30(6):513–528
8. Smirnova OA, Yonezawa M (2003) Radioprotection effect of low level preirradiation on mammals: modeling and experimental investigations. Health Phys 85(2):150–158
9. Smirnova OA, Yonezawa M (2004) Radioresistance in mammals induced by low-level chronic irradiation: modeling and experimental investigations. Health Phys 87(4):366–374
10. Hu S, Smirnova OA, Cucinotta FA (2012) A biomathematical model of lymphopoiesis following severe radiation accidents–potential use for dose assessment. Health Phys 102(4):425–436
11. Hu S, Blakely WF, Cucinotta FA (2014) HemoDose: a biodosimetry tool based on multi-type blood cell counts. Health Phys (in press)

12. Gregoire E, Hadjidekova V, Hristova R, Gruel G, Roch-Lefevre S, Voisin P, Staynova A, Deleva S, Ainsbury EA, Lloyd DC, Barquinero JF (2013) Biological dosimetry assessments of a serious radiation accident in Bulgaria in 2011. Radiat Prot Dosim 155(4):418–422
13. Djounova J, Guleva I, Negoicheva K, Mileva I, Panova D, Rupova I, Gigov I (2012) Initial medical diagnosis of patients severely irradiated in the accident with 60Co in Bulgaria. Radiat Prot Dosim 151(4):640–644
14. Costongs GM, Janson PC, Bas BM, Hermans J, van Wersch JW, Brombacher PJ (1985) Short-term and long-term intra-individual variations and critical differences of clinical chemical laboratory parameters. J Clin Chem Clin Biochem 23(1):7–16
15. Hirokawa K, Utsuyama M, Hayashi Y, Kitagawa M, Makinodan T, Fulop T (2013) Slower immune system aging in women versus men in the Japanese population. Immun Ageing 10(1):19
16. Ainsbury EA, Bakhanova E, Barquinero JF, Brai M, Chumak V, Correcher V, Darroudi F, Fattibene P, Gruel G, Guclu I, Horn S, Jaworska A, Kulka U, Lindholm C, Lloyd D, Longo A, Marrale M, Monteiro Gil O, Oestreicher U, Pajic J, Rakic B, Romm H, Trompier F, Veronese I, Voisin P, Vral A, Whitehouse CA, Wieser A, Woda C, Wojcik A, Rothkamm K (2011) Review of retrospective dosimetry techniques for external ionising radiation exposures. Radiat Prot Dosim 147(4):573–592
17. Djounova J, Guleva I, Negoicheva K, Mileva I, Panova D, Rupova I (2014) Clinical data from one year follow-up of victims of the radiation accident with 60Co in Bulgaria. Health Phys 107(3):248–254
18. Friesecke I, Beyrer K, Wedel R, Reimers K, Fliedner TM (2000) SEARCH: a system for evaluation and archiving of radiation accidents based on case histories. Radiat Environ Biophys 39(3):213–217
19. Gusev I, Guskova A, Mettler F (2001) Medical management of radiation accidents, 2nd edn. CRC Press, Boca Raton

FSC
www.fsc.org
MIX
Papier | Fördert gute Waldnutzung
FSC® C083411

Kolibri 360 GmbH
Ferdinand-Jühlke-Straße 7
99095 Erfurt, Deutschland
produktsicherheit@kolibri360.de